Contents

The Heritage Builders® Series

This resource was created as an outreach of the Heritage Builders Association—a network of families and churches committed to passing a strong heritage to the next generation. Designed to motivate and assist families as they become intentional about the heritage passing process, this series draws upon the collective wisdom of parents, grandparents, church leaders, and family life experts, in an effort to provide balanced, biblical parenting advice along with effective, practical tools for family living. For more information on the goals and work of Heritage Builders Association, please see page 103.

Kurt Bruner, M.A.
Executive Editor
Heritage Builders® Series

Introduction

There is toothpaste all over the plastic-covered table. Four young kids are having the time of their lives squeezing the paste out of the tube—trying to expunge every drop like Dad told them to. "Okay," says Dad, slapping a twenty-dollar bill onto the table. "The first person to get the toothpaste back into their tube gets this money!" Little hands begin working to shove the peppermint pile back into rolled-up tubes—with very limited success.

Jim is in the midst of a weekly routine in the Weidmann home when he and his wife spend time creating "impression points" with the kids. "We can't do it, Dad!" protests the youngest child.

"The Bible tells us that's just like your tongue. Once the words come out, it's impossible to get them back in. You need to be careful what you say because you may wish you could take it back." An unforgettable impression is made.

Impression points occur every day of our lives. Intentionally or not, we impress upon our children our values, preferences, beliefs, quirks, and concerns. It happens both through our talk and through our walk. When we do it right, we can turn them on to the things we believe. But when we do it wrong, we can turn them off to the values we most hope they will embrace. The goal is to find ways of making this reality work for us, rather than against us. How? By creating and capturing opportunities to impress upon the next generation our values and beliefs. In other words, through what we've labeled impression points.

The kids are all standing at the foot of the stairs. Jim is at the top of that same staircase. They wait eagerly for Dad's instructions.

"I'll take you to Baskin Robbins for ice cream if you can figure how to get up here." He has the attention of all four kids. "But there are a few rules. First, you can't touch the stairs. Second, you can't touch the railing. Now, begin!"

After several contemplative moments, the youngest speaks up. "That's impossible, Dad! How can we get to where you are without touching the stairs or the railing?"

After some disgruntled agreement from two of the other children, Jacob gets an idea. "Hey, Dad. Come down here." Jim walks down the stairs. "Now bend over while I get on your back. Okay, climb the stairs."

Bingo! Jim proceeds to parallel this simple game with how it is impossible to get to God on our own. But when we trust Christ's completed work on our behalf, we can get to heaven. A lasting impression is made. After a trip up the stairs on Dad's back, the whole gang piles into the minivan for a double scoop of mint-chip.

Six years ago, Jim and his wife Janet began setting aside time to intentionally impress upon the kids their values and beliefs through a weekly ritual called "family night." They play games, talk, study, and do the things which reinforce the importance of family and faith. It is during these times that they intentionally create these impression points with their kids. The impact? The kids are having fun and a heritage is being passed.

intentional or "oops"?

Sometimes, we accidentally impress the wrong things on our kids rather than intentionally impressing the right things. But there is an effective, easy way to change that. Routine family nights are a powerful tool for creating intentional impression points with our children.

The concept behind family nights is rooted in a biblical mandate summarized in Deuteronomy 6:5-9.

> *"Love the LORD your God with all your heart and with all your soul and with all your strength. These commandments that I give you today are to be upon your hearts. Impress them on your children."*
> ***How?***
> *"Talk about them when you sit at home and when you walk along the road, when you lie down and when you get up. Tie them as symbols on your hands and bind them on your foreheads. Write them on the doorframes of your houses and on your gates."*

In other words, we need to take advantage of every opportunity to impress our beliefs and values in the lives of our children. A growing network of parents are discovering family nights to be a highly effective, user-friendly approach to doing just that. As one father put it , "This has changed our entire family life." And another dad, "Our investment of time and energy into family nights has more eternal value than we may ever know." Why? Because they are intentionally teaching their children at the wisdom level, the level at which the children understand and can apply eternal truths.

truth is a treasure

Two boys are running all over the house, carefully following the complex and challenging instructions spelled out on the "truth treasure map" they received moments ago. An earlier map contained a few rather simple instructions that were much easier to follow. But the "false treasure box" it lead to left something to be desired. It was empty. Boo Dad! They hope for a better result with map number two.

STEP ONE:

Walk sixteen paces into the front family room.

STEP TWO:

Spin around seven times, then walk down the stairs.

STEP THREE:

Run backwards to the other side of the room.

STEP FOUR:

Try and get around Dad and climb under the table.

You get the picture. The boys are laughing at themselves, complaining to Dad, and having a ball. After twenty minutes of treasure hunting they finally reach the elusive "truth treasure box." Little hands open the lid, hoping for a better result this time around. They aren't disappointed. The box contains a nice selection of their favorite candies. Yea Dad!

"Which map was easier to follow?" Dad asks.

"The first one," comes their response.

"Which one was better?"

"The second one. It led to a true treasure," says the oldest.

"That's just like life," Dad shares, "Sometimes it's easier to follow what is false. But it is always better to seek and follow what is true."

They read from Proverbs 2 about the hidden treasure of God's truth and end their time repeating tonight's jingle—"It's best for you to seek what's true." Then they indulge themselves with a mouthful of delicious candy!

the power of family nights

The power of family nights is twofold. First, it creates a formal setting within which Dad and Mom can intentionally instill beliefs, values, or character qualities within their child. Rather than defer to the influence of peers and media, or abdicate character training to the school and church, parents create the opportunity to teach their children the things that matter most.

The second impact of family nights is perhaps even more significant than the first. Twenty to sixty minutes of formal fun and instruction can set up countless opportunities for informal reinforcement. These informal impression points do not have to be created, they just happen—at the dinner table, while driving in the car, while watching television, or any other parent/child time together. Once you have formally discussed a given family night topic, you and your children will naturally refer back to those principles during the routine dialogues of everyday life.

If the truth were known, many of us hated family devotions while growing up. We had them sporadically at best, usually whenever our parents were feeling particularly guilty. But that was fine, since the only thing worse was a trip to the dentist. Honestly, do we really think that is what God had in mind when He instructed us to teach our children? As an alternative, many parents are discovering family nights to be a wonderful complement to or replacement for family devotions as a means of passing their beliefs and values to the kids. In fact, many parents hear their kids ask at least three times per week:

"Can we have family night tonight?"

Music to Dad's and Mom's ears!

Keys to Effective Family Nights

There are several keys which should be incorporated into effective family nights.

MAKE IT FUN!

Enjoy yourself, and let the kids have a ball. They may not remember everything you say, but they will always cherish the times of laughter—and so will you.

KEEP IT SIMPLE!

The minute you become sophisticated or complicated, you've missed the whole point. Don't try to create deeply profound lessons. Just try to reinforce your values and beliefs in a simple, easy-to-understand manner. Read short passages, not long, drawn-out sections of Scripture. Remember: The goal is to keep it simple.

DON'T DOMINATE!

You want to pull them into the discovery process as much as possible. If you do all the talking, you've missed the mark. Ask questions, give assignments, invite participation in every way possible. They will learn more when you involve all of their senses and emotions.

GO WITH THE FLOW!

It's fine to start with a well-defined outline, but don't kill spontaneity by becoming overly structured. If an incident or question leads you in a different direction, great! Some of the best impression opportunities are completely unplanned and unexpected.

MIX IT UP!

Don't allow yourself to get into a rut or routine. Keep the sense of excitement and anticipation through variety. Experiment to discover what works best for your family. Use books, games, videos, props, made-up stories, songs, music or music videos, or even go on a family outing.

DO IT OFTEN!

We tend to find time for the things that are really important. It is best to set aside one evening per week (the same evening if possible) for family night. Remember, repetition is the best teacher. The more impressions you can create, the more of an impact you will make.

MAKE A MEMORY!

Find ways to make the lesson stick. For example, just as advertisers create "jingles" to help us remember their products, it is helpful to create family night "jingles" to remember the main theme—such as "It's best for you to seek what's true" or "Just like air, God is there!"

USE OTHER TOOLS FROM THE HERITAGE BUILDERS TOOL CHEST!

Family night is only one exciting way for you to intentionally build a loving heritage for your family. You'll also want to use these other exciting tools from Heritage Builders.

The Family Fragrance: There are five key qualities to a healthy family fragrance, each contributing to an environment of love in the home. It's easy to remember the Fragrance Five by fitting them into an acrostic using the word "Aroma"—

A—Affection
R—Respect
O—Order
M—Merriment
A—Affirmation

Impression Points: Ways that we impress on our children our values, preferences, and concerns. We do it through our talk and our actions. We do it intentionally (through such methods as Family Nights), and we do it incidentally.

The Right Angle: The Right Angle is the standard of normal healthy living against which our children will be able to measure their attitudes, actions, and beliefs.

Traditions: Meaningful activities which the process of passing on emotional, spiritual, and relational inheritance between generations. Family traditions can play a vital role in this process.

Please see the back of the book for information on how to receive the FREE Heritage Builders Newsletter which contains more information about these exciting tools! Also, look for the new book, *The Heritage,* available at your local Christian bookstore.

How to Use This Tool Chest

Summary page: For those who like the bottom line, we have provided a summary sheet at the start of each family night session. This abbreviated version of the topic briefly highlights the goal, key Scriptures, activity overview, main points, and life slogan. On the reverse side of this detachable page there is space provided for you to write down any ideas you wish to add or alter as you make the lesson your own.

Step-by-step: For those seeking suggestions and directions for each step in the family night process, we have provided a section which walks you through every activity, question, Scripture reading, and discussion point. Feel free to follow each step as written as you conduct the session, or read through this portion in preparation for your time together.

À la carte: We strongly encourage you to use the material in this book in an "à la carte" manner. In other words, pick and choose the questions, activities, Scriptures, age-appropriate ideas, etc. which best fit your family. This book is not intended to serve as a curriculum, requiring compliance with our sequence and plan, but rather as a tool chest from which you can grab what works for you and which can be altered to fit your family situation.

The long and the short of it: Each family night topic presented in this book includes several activities, related Scriptures, and possible discussion items. Do not feel it is necessary to conduct them all in a single family night. You may wish to spread one topic over several weeks using smaller portions of each chapter, depending upon the attention span of the kids and the energy level of the parents. Remember, short and effective is better than long and thorough.

Journaling: Finally, we have provided space with each session for you to capture a record of meaningful comments, funny happenings, and unplanned moments which will inevitably occur during family night. Keep a notebook of these journal entries for future reference. You will treasure this permanent record of the heritage passing process for years to come.

1: Our Protector

Exploring how God protects us

Scripture
- Matthew 6:26-27
- 2 Thessalonians 3:3
- Psalm 18:2-3

ACTIVITY OVERVIEW		
Activity	**Summary**	**Pre-Session Prep**
Activity 1: A Crushing Eggsperience	Attempt to crush a raw egg and learn about God's protection.	You'll need one or two raw eggs, a sink or bucket, and a Bible.
Activity 2: God's Strength	Place a flame under two different balloons and learn about how God strengthens us.	You'll need two un-inflated black balloons, water, a candle, matches, and a Bible.

Main Points:

—God loves us and protects us.

—God strengthens us and protects us from Satan.

LIFE SLOGAN: "When with God we connect, He will always protect."

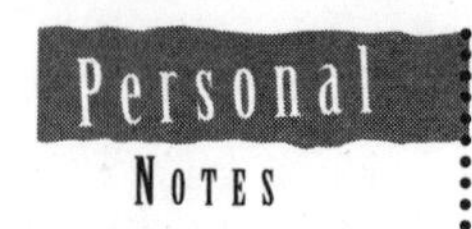

Make it your own
In the space provided below, outline the flow and add any additional ideas to guide you through the process of conducting this family night.

Prayer & Praise Items
In the space provided below, list any items you wish to pray about or give praise for during this family night session.

Journal
In the space provided below, capture a record of any fun or meaningful things which happened during this family night session.

Session Tip

We intentionally have provided more material than we would expect to be used in a single "Family Night" session. You know your family's unique interests and life circumstances best, so feel free to adapt this lesson to meet your family members' needs. Remember, short and simple is better than long and comprehensive.

WARM-UP

Open with Prayer: Begin by having a family member pray, asking God to help everyone in the family understand more about Him through this time. After prayer, review your last lesson by asking these questions:

- **What did we learn about in our last lesson?**
- **What was the Life Slogan?**
- **Have your actions changed because of what we learned? If so, how?** Encourage family members to give specific examples of how they've applied learning from the past week.

Share: Today we're going to have fun as we explore how God protects and strengthens us.

ACTIVITY 1: A Crushing Eggsperience

Point: God loves and protects us.

Supplies: You'll need one or two raw eggs, a sink or bucket, and a Bible.

Activity: Begin by illustrating the fragile nature of an egg. Take family members over to a sink and carefully crack an egg by dropping it or hitting it against the side of the sink. (You could also do this over a bucket or outside where cleanup will be easier.) Pass the eggshell around, and ask:

- **What do you notice about this shell?** (It's fragile; it's light; there's not much to it.)

Ask for a volunteer who thinks he or she can crush the egg with the "raw" power of their hand. Then place another raw egg in a child's hand so that the ends are at the thumb and pinky (see illustration). Help the child carefully wrap his or her hand around the egg and attempt to crush it. In most cases, this will be impossible, as the pressure is distributed equally around the shell. At the very least, it will be a challenge for a child to crush the egg. Have all family members attempt this activity.

Discuss the following questions:

- **What was it like to try and crush the egg?** (It wasn't easy; I couldn't do it.)
- **What does this tell you about the eggshell?** (It is stronger than it looks; it protects the growing baby chick.)

Read aloud Matthew 6:26-27, then consider these questions:

- **If God shows us how much He protects His creation, how does God show His love and protection for us?** (By helping us make good choices; by giving us friends who care for us.)
- **How is the way God protects the chicks with the eggshell like the way He protects us?** (He puts a "shell" around us; He knows just how strong to make the covering.)

Age Adjustments

OLDER CHILDREN AND TEENAGERS may find the egg-crushing experience quite surprising—more so than younger children. Take advantage of this teachable moment to help them realize God's marvelous creativity in creation—and to discuss how God knows just what our needs are and provides for them in innovative and surprising ways.

Share: God will never leave us and knows what's best for us. He's always there to listen and can meet all our needs. He knows just what kind of protection we need, and cares for us.

ACTIVITY 2: God's Strength

Point: God strengthens us and protects us from Satan.

Supplies: You'll need two un-inflated black balloons, water, a candle, matches, and a Bible.

Activity: Begin the activity by asking family members to tell about times they've been worried or scared. Ask:

- **What did it feel like to be worried or scared?**
- **What were you afraid of?**
- **Did you think about God when you were feeling this way? Why or why not?**

Read aloud 2 Thessalonians 3:3 and Psalm 18:2-3. Then consider this question:

- **What do these passages tell us about God's protection?** (God gives us strength; God cares for us.)

Share: God is the source of our strength—He's our rock. He is like a shield who protects us from our enemies and especially from Satan.

NOTE: You'll want to practice this activity before doing it with your family to determine the right distance to hold the candle . . . and the right amount of water to place in the balloon.

Fill one of the black balloons with enough water to cover the bottom of the balloon when blown up. Blow up both balloons to the same size. Light the candle.

Ask family members what they think will happen when you place the balloons above the flame. Then hold the empty balloon over the flame and watch it pop.

 Ask:

- **Did anything protect the balloon? Why not?** (The fire was too strong; it was too thin.)

Hold the other balloon above the flame. Watch it carefully and it won't pop. After you blow out the candle, toss the balloon around to family members and discuss the following:

- **Why didn't the balloon pop?** (It had water in it; it was protected.)
- **How is this balloon like people who love Jesus?** (They are protected; they have something inside that keeps them safe.)

Share: When we love Jesus, we are protected from Satan just as this balloon was protected from the flame. God strengthens us to withstand Satan's attempts to "pop" us. We can be thankful that God lives in us and protects us.

WRAP-UP

Gather everyone in a circle and have family members take turns answering this question: **What's one thing you've learned about God today?**

Next, tell kids you've got a new "Life Slogan" you'd like to share with them.

Life Slogan: Today's Life Slogan is this: "When with God we connect, He will always protect." Have family members repeat the slogan two or three times to help them learn it. Then encourage them to practice saying it during the week so they can talk about it at your next family night session.

Close in Prayer: Allow time for each family member to share prayer concerns and answers to prayer. Then close your time together with prayer for each concern. Thank God for listening to and caring about us.

Remember to record your prayer requests so you can refer to them in the future as you see God answering them.

2: Peer Pressure

Exploring how God can help us face negative peer pressure

Scripture
- Matthew 14:6-12
- Luke 23:13-25
- Proverbs 1:8-10
- Proverbs 12:3

ACTIVITY OVERVIEW

Activity	Summary	Pre-Session Prep
Activity 1: Giving in to Peer Pressure	Watch as an eye-dropper is "pressured" to sink.	You'll need an empty 2-liter plastic bottle, an eyedropper, water, and a Bible.
Activity 2: Pressure in a Jar	Attempt to position a string so it falls into a jar and learn about the power of peer pressure.	You'll need a jar, string, a chair, a fan, a small weight, and a Bible.

Main Points:

—Do not give in to those around you.
—Stand strong in the Lord.

LIFE SLOGAN: "When you feel the pressure, don't sink; focus on God and think!"

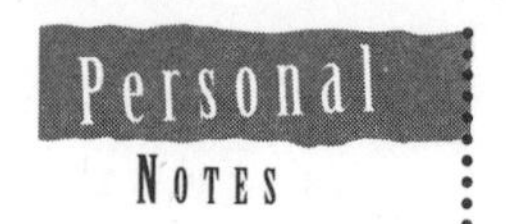

Make it your own

In the space provided below, outline the flow and add any additional ideas to guide you through the process of conducting this family night.

Prayer & Praise Items

In the space provided below, list any items you wish to pray about or give praise for during this family night session.

Journal

In the space provided below, capture a record of any fun or meaningful things which happened during this family night session.

Session Tip

We intentionally have provided more material than we would expect to be used in a single "Family Night" session. You know your family's unique interests and life circumstances best, so feel free to adapt this lesson to meet your family members' needs. Remember, short and simple is better than long and comprehensive.

WARM-UP

Open with Prayer: Begin by having a family member pray, asking God to help everyone in the family understand more about Him through this time. After prayer, review your last lesson by asking these questions:

- **What did we learn about in our last lesson?**
- **What was the Life Slogan?**
- **Have your actions changed because of what we learned? If so, how?** Encourage family members to give specific examples of how they've applied learning from the past week.

Share: Today we're going to learn about negative peer pressure—and how to face it and win with God's help.

ACTIVITY 1: Giving in to Peer Pressure

Point: Do not give in to those around you.

Supplies: You'll need an empty 2-liter plastic bottle, an eyedropper, water, and a Bible.

Activity: Read Matthew 14:6-12 and Luke 23:13-25. Then discuss the following questions:

- **What are these passages about?** (People who did wrong things.)
- **What things are the same about these two stories?** (Both people knew what they were doing was wrong.)
- **What things made them do the wrong things?** (People forced them to change; friends convinced them to do what was wrong.)

Share: The Bible tells us that King Herod was "distressed" and that Pilate had wanted to release Jesus: both men knew what they were doing was wrong, but they gave in and did it anyway. That's what we

call negative peer pressure.

Have family members each share about a time they were in a similar situation—they felt the pressure of friends or other people to do something they knew was wrong.

NOTE: You may want to test this before doing the activity with your family. Fill the 2-liter bottle with water. Then take the eyedropper and fill it about half full of water and drop it into the bottle. If it sinks, let some of the water out of the dropper and try again. You'll want just enough water in the dropper so it floats. Put the cap on the bottle.

Give the bottle a gentle squeeze. As you squeeze and put pressure on the bottle, the eyedropper will sink. If it doesn't, or if you must squeeze very hard to get it to sink, open the bottle and add just a little more water to the dropper. This can be a little tricky, but the result is worth the effort.

When the dropper is working correctly, have family members each take a turn squeezing the bottle and watching it sink and float. Then consider the following questions:

- **What happens when the bottle is squeezed?** (The dropper falls; the dropper sinks.)
- **How is this like what happens to the people in the Bible stories when they're pressured by others?** (They fall; they sink and do something wrong.)
- **How is this like the way you feel when others pressure you?** (I feel down; I am unhappy when others pressure me.)

Share: People will pressure us to do the wrong thing, but God can give us the strength to stay strong and not "sink" to their level.

ACTIVITY 2: Pressure in a Jar

Point: Stand strong in the Lord.

Supplies: You'll need a jar, string, a chair, a fan, a small weight, and a Bible.

Activity: Set a jar on the floor and place a fan nearby, facing the jar. Set a chair next to the jar as well. Have a volunteer stand on the chair. Give your volunteer a long string (long enough to reach the floor) and have him or her attempt to drop the hanging end of the string into the jar without bending over. Give each family member a chance to successfully complete this simple exercise.

 Ask:

- **How easy or difficult was it to drop the string into the jar?** (It wasn't too bad; it was easy.)

Now turn on the fan so it blows toward the string. Have family members repeat the activity again. Make sure the fan blows strong enough to keep the string from hanging straight. Afterward, consider these questions:

- **How was this activity different from the first time?** (It was impossible; it was much harder to drop the string.)
- **How is the fan's power like the power of friends to sway or change your mind?** (Friends try to tell you what to do; when others try to convince you of something, it's harder to do what's right.)

Read Proverbs 1:8-10. Then **share: When we try to live our lives by doing the right thing, it's a challenge (just like the first time we tried to drop the string). But when others are trying to distract us or change our minds, it's even more difficult. God wants us to stay away from people who pressure us to do the wrong thing.**

Attach a weight to the bottom of the string and repeat the activity with the fan blowing.

Read or explain Proverbs 12:3. Then **share: When we trust God to help us, and do our best to do what's right, it's easier to do the right thing, even when others pressure us to do something wrong. The weight is like our faith in God. When we are righteous, we can withstand the winds and pressures to do what's wrong.**

 Ask:

- **What do the Bible passages and these activities tell us about facing peer pressure?** (When we trust God, we can beat

Age Adjustments

YOUNGER CHILDREN may have difficulty dropping the string into the jar. You may need to help steady their arms to make it easier to complete the task. Or, you may want to choose something more substantial than string—such as yarn—to do this activity. Older children and teenagers might enjoy the greater challenge of attempting this from a higher starting place such as a stairway or the top of a bunk bed.

the pressure; when we allow others to sway us, it's not easy to do the right thing; we need to do what's right, even when others want us to do something wrong.)

WRAP-UP

Gather everyone in a circle and have family members take turns answering this question: **What's one thing you've learned about God today?**

Next, tell kids you've got a new "Life Slogan" you'd like to share with them.

Life Slogan: Today's Life Slogan is this: "When you feel the pressure, don't sink; focus on God and think!" Have family members repeat the slogan two or three times to help them learn it. Then encourage them to practice saying it during the week so they can talk about it at your next family night session.

Close in Prayer: Allow time for each family member to share prayer concerns and answers to prayer. Then close your time together with prayer for each concern. Thank God for listening to and caring about us.

Remember to record your prayer requests so you can refer to them in the future as you see God answering them.

3: Dependence on God

Exploring how dependency is a position of power

Scripture
- Ephesians 6:16
- 1 Kings 11:3-4
- 2 Corinthians 12:9-10

ACTIVITY OVERVIEW		
Activity	**Summary**	**Pre-Session Prep**
Activity 1: Fiery Darts	Watch as a skewer is inserted into a balloon without popping the balloon, and talk about Satan's attacks on us.	You'll need balloons, long darts or shish kebab skewers, cooking oil, and a Bible.
Activity 2: An Eye for Weaknesses	Discover which eye is dominant and learn about weaknesses.	You'll need two pieces of plain white paper, a pencil, and a Bible.

Main Points:

—We must hold firm to our faith and depend on God for our strength.

—We all have weaknesses and will be attacked by Satan.

LIFE SLOGAN: "I will depend on God for power; each and every hour."

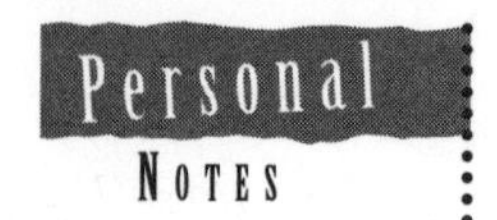

Make it your own
In the space provided below, outline the flow and add any additional ideas to guide you through the process of conducting this family night.

Prayer & Praise Items
In the space provided below, list any items you wish to pray about or give praise for during this family night session.

Journal
In the space provided below, capture a record of any fun or meaningful things which happened during this family night session.

Session Tip

We intentionally have provided more material than we would expect to be used in a single "Family Night" session. You know your family's unique interests and life circumstances best, so feel free to adapt this lesson to meet your family members' needs. Remember, short and simple is better than long and comprehensive.

WARM-UP

Open with Prayer: Begin by having a family member pray, asking God to help everyone in the family understand more about Him through this time. After prayer, review your last lesson by asking these questions:

- **What did we learn about in our last lesson?**
- **What was the Life Slogan?**
- **Have your actions changed because of what we learned? If so, how?** Encourage family members to give specific examples of how they've applied learning from the past week.

Share: Today we'll explore how Satan attempts to hurt us, and why it's important to depend on God for our strength.

ACTIVITY 1: Fiery Darts

Point: We must hold firm to our faith and depend on God for our strength.

Supplies: You'll need balloons, long darts or shish kebab skewers, cooking oil, and a Bible.

Activity: Blow a balloon up halfway and explain that the balloon represents us. Ask people to share what they think the air would represent, then suggest that just as the balloon holds air, we try to live our lives full of God's love and doing His will.

Then hold up a dart or skewer and ask:

- **What do you think will happen when I poke the balloon with this dart?** (It will pop.)

Go ahead and pop the balloon.

Read aloud Ephesians 6:16 and then **share: Satan is trying to destroy our relationship with Christ by throwing temptations our way. And just as the dart popped the balloon, our relationship is damaged when we give in to those fiery darts from Satan.**

Ask:

- **How does Satan attack us?** (By tempting us; by making us doubt God; by telling lies.)
- **What are some of the ways your relationship with God has been "popped"?** (I've lied; I have been mean to someone; I haven't prayed.)

Now blow up another balloon halfway and ask family members once again what they think will happen when you poke the balloon with the dart or skewer. This time, lightly oil the dart and carefully push it slowly through the nipple end of the balloon. (You may wish to practice this ahead of time to assure success.) The balloon should not pop as long as the dart or skewer is left in the balloon during your discussion and explanation.

Consider this question:

- **What did you think would happen this time?** (The balloon would still pop.)

Share: When I oiled the dart and chose where to poke it on the balloon, it didn't pop the balloon. I prepared the balloon for the sharp dart and kept it from harm. Just as I prepared this balloon, we need to be prepared for the attacks we'll face in life. When we trust God fully, we will become stronger and better able to keep doing God's will even when Satan shoots his fiery darts our way.

Age Adjustments

OLDER CHILDREN AND TEENAGERS can go deeper with this activity by discussing the difference between Satan's "attacks" and the popular phrase "the devil made me do it." Discuss with older children and teenagers how Satan's influence is certainly felt . . . but that we're responsible for how we respond to that influence. Help your children discover that with God's help, they can fend off attacks and not feel like a victim of Satan's ploys.

ACTIVITY 2: An Eye for Weaknesses

Point: We all have weaknesses and will be attacked by Satan.

Supplies: You'll need two pieces of plain white paper, a pencil, and a Bible.

Activity: Poke a pencil-sized hole through the middle of one piece of paper. In the middle of the other paper, draw and color in a circle the size of a penny.

Have your family members complete the following tasks to determine which eye is dominant:

1. Hold the paper with the colored circle at arm's length.

2. Hold the paper with the hole in it in front at just the distance where the circle can be seen by both eyes (through the hole).

3. Alternate closing the left and right eye to determine when the circle "disappears."

4. If the circle disappears when the right eye is closed, that means the person is right-eye dominant.

After everyone has completed this activity, explain: **The reason the dot went away when you closed one of your eyes is because each of us has one eye that is dominant or stronger than the other. We also have strong and weak spots in other areas of our lives.**

Read 1 Kings 11:3-4. Then **share: Even the wisest man who ever lived, Solomon, had weak spots. He could not say no to his wives who led him to do some wrong things. Satan likes to attack our weak spots.**

Have family members each share one or more weak spots in their lives, such as lying, cheating, gossiping, selfishness, lust, or greed. It's a good idea to share one of your own weak spots to illustrate to children that even adults are vulnerable to Satan's attacks.

Then **read** aloud 2 Corinthians 12:9-10.

Ask:

- **What does this passage tell us about what to do with our weak spots?** (Admit we have them; tell someone what they are; be honest.)
- **How do we overcome our weak spots?** (Ask God to help; trust God.)

Share: We all have weaknesses and "blind spots" where Satan will attack. If we have a weak spot telling the truth, we'll be given lots of temptation to lie. But with the power of the Holy Spirit in us, we can overcome our weaknesses and strengthen our lives to be able to do the right thing.

Age Adjustments

Younger Children may have some difficulty understanding this activity. You may want to consider a simpler exploration of strength versus weakness by having children attempt to lift an object using only one finger— first their "pointer," and then their "ring finger." Children will likely find that their pointers are much stronger. You can use this illustration to jump into the discussion of how we need to trust God to help us with our weaknesses.

WRAP-UP

Gather everyone in a circle and have family members take turns answering this question: **What's one thing you've learned about God today?**

Next, tell kids you've got a new "Life Slogan" you'd like to share with them.

Life Slogan: Today's Life Slogan is this: "I will depend on God for power; each and every hour." Have family members repeat the slogan two or three times to help them learn it. Then encourage them to practice saying it during the week so they can talk about it at your next family night session.

Close in Prayer: Allow time for each family member to share prayer concerns and answers to prayer. Then close your time together with prayer for each concern. Thank God for listening to and caring about us.

Remember to record your prayer requests so you can refer to them in the future as you see God answering them.

4: The Power of God

Exploring how nothing is impossible when it is in God's will

Scripture
- Isaiah 40:29-31
- Matthew 21:18-22

ACTIVITY OVERVIEW

Activity	Summary	Pre-Session Prep
Activity 1: One Strong Potato	Watch a straw being plunged through a potato, and learn about the source of our strength: God.	You'll need straws, fresh baking potatoes, and a Bible.
Activity 2: Nothing Is Impossible	See how a hard-boiled egg can be sucked into a glass bottle and discover that nothing is impossible with God.	You'll need a hard-boiled egg, butter, a glass bottle with a mouth just a little smaller than the egg, paper, matches, and a Bible.

Main Points:

—God is our only source of strength.
—Nothing is impossible when it is in God's will.

LIFE SLOGAN: "My strength comes from Him; not from within."

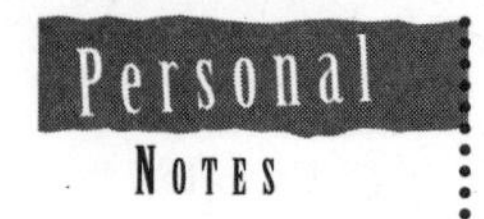

Make it your own
In the space provided below, outline the flow and add any additional ideas to guide you through the process of conducting this family night.

Prayer & Praise Items
In the space provided below, list any items you wish to pray about or give praise for during this family night session.

Journal
In the space provided below, capture a record of any fun or meaningful things which happened during this family night session.

Session Tip

We intentionally have provided more material than we would expect to be used in a single "Family Night" session. You know your family's unique interests and life circumstances best, so feel free to adapt this lesson to meet your family members' needs. Remember, short and simple is better than long and comprehensive.

WARM-UP

Open with Prayer: Begin by having a family member pray, asking God to help everyone in the family understand more about Him through this time. After prayer, review your last lesson by asking these questions:

- **What did we learn about in our last lesson?**
- **What was the Life Slogan?**
- **Have your actions changed because of what we learned? If so, how?** Encourage family members to give specific examples of how they've applied learning from the past week.

Share: Today we're going to discover the source of our strength—and be surprised at the things God can do.

ACTIVITY 1: One Strong Potato

Point: God is our only source of strength.

Supplies: You'll need straws, fresh baking potatoes, and a Bible.

Activity: Ask family members to recall Bible stories where God helped overcome situations that seemed hopeless or impossible. Here are a few to get you started: God parts the Red Sea for Moses and the Israelites; God topples the walls of Jericho; Jesus feeds the 5,000 with just a small amount of food; Jesus raises Lazarus from the dead.

After talking about these situations, hold up a potato and a straw.

Ask:

- **How can I get this straw through the potato without bending it?** (Drill a hole in the potato; you can't.)

Discuss how this activity might seem impossible, then grip the straw by pinching one end with your thumb and forefinger and plunge it into the potato at a right angle to the surface. You may want to practice this a few times to be sure you know how to do it. It will take the proper angle and amount of force for this to succeed.

NOTE: If you can't find a fresh baking potato to do this with, you can use one that's been soaked in water for an hour.

Read aloud Isaiah 40:29-31.
Then **share: Most people wouldn't think it is possible to poke this straw through a potato. But it is possible as we can see. Sometimes, we face things that are bigger than we are—things that seem impossible like crossing the Red Sea. And during those times, we forget to consider God's role in that situation, just as we didn't consider the power of the straw to go through the potato.**

Have family members attempt to plunge the straw through the potato. They'll soon discover that the only way it works is when the angle and force are just right. Explain that this is also how it is when it comes to situations we face that seem impossible: when we're in alignment with God's will—when we're trusting Him and accepting His guidance—He can penetrate our situation and help make things work out.

Age Adjustments

HELP YOUNGER CHILDREN discover that God is bigger than our biggest problems with this simple exercise. Have children share about a problem they think is really big (such as having trouble with school, being afraid of the dark, and so on). Tell them to imagine that problem is as big as their bed. Go to their bedroom and have them sit on their bed to imagine how it must feel to face such a big problem. Then explain that God's power is bigger than your house, your neighborhood, your city, and your world—much bigger than their bed. This will help younger children discover the possibility of God's intervention by giving them a sense of perspective on God's great power.

ACTIVITY 2: Nothing Is Impossible

Point: Nothing is impossible when it is in God's will.

Supplies: You'll need a hard-boiled egg, butter, a glass bottle with a mouth just a little smaller than the egg, paper, matches, and a Bible.

Activity: Set a hard-boiled egg (with the shell peeled off) on top of the bottle and ask children to get the egg into the bottle by simply telling it to go in. Obviously, this won't work and children will find it somewhat ridiculous to attempt it. That's OK. Enjoy the silliness of this before continuing.

Have children share ideas on how the egg could be put into the bottle without breaking either the bottle or the egg. Then lightly grease the jar's mouth using the butter, hold the bottle sideways, and slowly slide a lighted piece of paper into it. Rotate the bottle upright and place it on the table. Place the egg vertically on the bottle (so it seals the opening). The fire will create a vacuum as it uses up the oxygen and will suck the egg into the bottle whole!

Read aloud Matthew 21:18-22. Then consider the following question:

- **How did the disciples react when He said that the tree would not bear fruit again?** (Surprised; they didn't believe Him.)

Share: When I asked you how to get the egg into the bottle whole, you couldn't do it by telling it. You may have thought it was impossible, just as the disciples thought what Jesus had done was impossible. But with God, all things are possible.

Ask children if they think you can remove the egg from the jar whole. Encourage them to believe in the possibility just as they should believe that God can do seemingly impossible things. Then show them how to get the egg out by turning the bottle upside-down so the egg creates a one-way valve sealing the mouth from the inside. Blow into the bottle as long and hard as you can, then firmly (without breaking the bottle) set the bottle right-side up on a table. The egg should pop out of the top.

For the next few weeks, each time you enjoy eggs for breakfast, remind family members of this activity and that nothing is impossible with God.

WRAP-UP

Gather everyone in a circle and have family members take turns answering this question: **What's one thing you've learned about God today?**

Next, tell kids you've got a new "Life Slogan" you'd like to share with them.

Life Slogan: Today's Life Slogan is this: "My strength comes from Him; not from within." Have family members repeat the slogan two or three times to help them learn it. Then encourage them to practice saying it during the week so they can talk about it at your next family night session.

Close in Prayer: Allow time for each family member to share prayer concerns and answers to prayer. Then close your time together with prayer for each concern. Thank God for listening to and caring about us.

Remember to record your prayer requests so you can refer to them in the future as you see God answering them.

5: A Changed Life

Exploring how we can be lights in a dark world

Scripture
- John 3:5-8
- Matthew 5:14-16
- 1 Timothy 2:1-4

ACTIVITY OVERVIEW		
Activity	**Summary**	**Pre-Session Prep**
Activity 1: Hair-Dryer Science	Learn about the role of the Holy Spirit.	You'll need 1/4-full roll of toilet paper, a blow dryer, a dowel rod, and a Bible.
Activity 2: Light in the Dark	Make candies "spark" and learn how to be a light in the world.	You'll need Wintergreen LifeSavers (or other similar candies), and a Bible.

Main Points:

—When we accept Jesus' gift of salvation, we receive the Holy Spirit.

—The Holy Spirit helps us to be a light in a dark world.

LIFE SLOGAN: "With God's might; you can be a light."

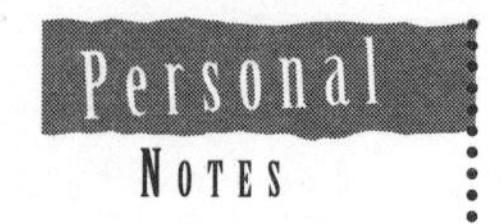

Make it your own

In the space provided below, outline the flow and add any additional ideas to guide you through the process of conducting this family night.

Prayer & Praise Items

In the space provided below, list any items you wish to pray about or give praise for during this family night session.

Journal

In the space provided below, capture a record of any fun or meaningful things which happened during this family night session.

Session Tip

We intentionally have provided more material than we would expect to be used in a single "Family Night" session. You know your family's unique interests and life circumstances best, so feel free to adapt this lesson to meet your family members' needs. Remember, short and simple is better than long and comprehensive.

WARM-UP

Open with Prayer: Begin by having a family member pray, asking God to help everyone in the family understand more about Him through this time. After prayer, review your last lesson by asking these questions:

- **What did we learn about in our last lesson?**
- **What was the Life Slogan?**
- **Have your actions changed because of what we learned? If so, how?** Encourage family members to give specific examples of how they've applied learning from the past week.

Share: Today we're going to learn how the Holy Spirit helps us to be light to a dark world.

ACTIVITY 1: Hair-Dryer Science

Point: When we accept Jesus' gift of salvation, we receive the Holy Spirit.

Supplies: You'll need 1/4-full roll of toilet paper, a blow dryer, a dowel rod, and a Bible.

Activity: Place the 1/4 roll of toilet paper on a dowel rod so the paper is positioned to unroll from the top side facing away from you. Have a volunteer turn on the hair dryer and slowly lower its air stream until it is just above the top of the toilet paper roll. The paper will begin to unroll without a direct flow of air. (For those of you who are interested in science, this is an example of Bernoulli's Principle which answers the question "why do airplanes fly?")

After enjoying the flying toilet paper, gather family members together and discuss the following questions:

- **What caused the paper to unroll?** (The air from the hair dryer.)
- **How do you know the air caused it? Could you see it?** (I just know; no, I couldn't see it, but I believe that's what did it.)

Read aloud John 3:5-8 and then have family members summarize what Jesus said about the wind. Then **share: When we accept Jesus as our Savior, we are given an invisible gift—the Holy Spirit! We cannot see the Holy Spirit, just as we can't see the wind and couldn't see the air from the hair dryer. But just as the air unrolled the toilet paper, the Holy Spirit works in ways we don't understand. And just as we saw the effects of the air on the paper roll, we can see the effects of the Holy Spirit in our lives.**

Have family members share ways they can tell the Holy Spirit is in someone's life. For example, someone might say: "I know the Holy Spirit is in me because I know how to do the right thing" or "I know the Holy Spirit is in me because I feel God's presence."

Age Adjustments

Older Children and Teenagers might enjoy a little "side trip" into a discussion of Bernoulli's Principle. A quick trip to the library or a reference web site should give you the material you need to delve deeper into this topic. Kids will be fascinated by the principle of flight. Take advantage of this sense of wonder to help children discover the wonderful creativity of God's Creation.

ACTIVITY 2: Light in the Dark

Point: The Holy Spirit helps us to be a light in a dark world.

Supplies: You'll need Wintergreen or Cryst-O-Mint LifeSavers (or other similar candies), and a Bible.

Activity: Read together Matthew 5:14-16.

Consider these questions:

- **What does it mean to be a light in the world?** (To do good things that others will see; to act like Jesus.)
- **How easy or difficult is it to be a light?** (It's hard sometimes when other people pressure us to do something wrong; it's easy, because I want to do the right thing.)

Have family members share about when it's been difficult to "let their light shine." Then **read** aloud 1 Timothy 2:1-4 and help family members discover that prayer helps us to live lives that are godly and holy—lives that allow us to shine in the darkness.

Take time for "popcorn prayers" with your family. Have family members each pray brief prayers, asking for God's help with a particular struggle. Then give everyone a LifeSaver and go together into a dark room (such as a closet or a bathroom). Have family members face each other and chew their LifeSavers with their mouths open (manners will have to be set aside for this activity!). Watch as the "sparks" come out of their mouths from the crunching of the candies. As you enjoy the spray of sparks, pray for your family to continue to be lights in a dark world—to spark with God's love so others can see.

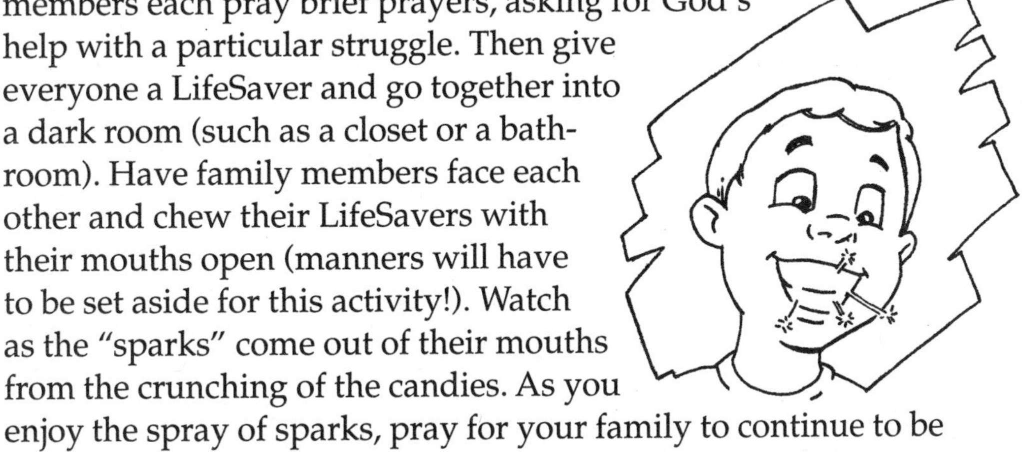

WRAP-UP

Gather everyone in a circle and have family members take turns answering this question: **What's one thing you've learned about God today?**

Next, tell kids you've got a new "Life Slogan" you'd like to share with them.

Life Slogan: Today's Life Slogan is this: "With God's might; you can be a light." Have family members repeat the slogan two or three times to help them learn it. Then encourage them to practice saying it during the week so they can talk about it at your next family night session.

Close in Prayer: Allow time for each family member to share prayer concerns and answers to prayer. Then close your time together with prayer for each concern. Thank God for listening to and caring about us.

Remember to record your prayer requests so you can refer to them in the future as you see God answering them.

6: To Know Is to Grow

Exploring ways to grow closer to God

Scripture
- Acts 9:1-18
- Psalm 119:105
- 2 Chronicles 34:31
- Acts 17:11
- James 1:22-25

ACTIVITY OVERVIEW		
Activity	**Summary**	**Pre-Session Prep**
Activity 1: Continuous Change	Taste sweetened and unsweetened lemonade and learn how the Holy Spirit makes us more like Jesus.	You'll need a pitcher, lemonade mix (sugarless), sugar, dry ice, and a Bible.
Activity 2: The Raisin for Bible Study	Watch "dancing raisins" and discover the need for daily Bible study.	You'll need raisins, a clear drinking glass, a 2-liter bottle of a clear soft drink (such as 7-Up or Sprite), and a Bible.

Main Points:

—We need to grow closer to Jesus each day.

—We need to feed on God's Word to grow in Christ.

LIFE SLOGAN: "The Word and Spirit are our link; without them we would surely sink."

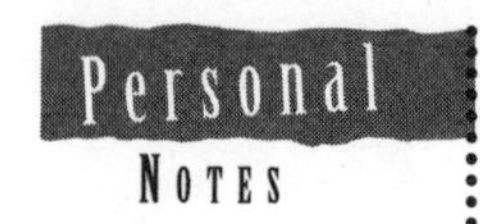

Make it your own

In the space provided below, outline the flow and add any additional ideas to guide you through the process of conducting this family night.

Prayer & Praise Items

In the space provided below, list any items you wish to pray about or give praise for during this family night session.

Journal

In the space provided below, capture a record of any fun or meaningful things which happened during this family night session.

Session Tip

We intentionally have provided more material than we would expect to be used in a single "Family Night" session. You know your family's unique interests and life circumstances best, so feel free to adapt this lesson to meet your family members' needs. Remember, short and simple is better than long and comprehensive.

WARM-UP

Open with Prayer: Begin by having a family member pray, asking God to help everyone in the family understand more about Him through this time. After prayer, review your last lesson by asking these questions:

- **What did we learn about in our last lesson?**
- **What was the Life Slogan?**
- **Have your actions changed because of what we learned? If so, how?** Encourage family members to give specific examples of how they've applied learning from the past week.

Share: Today we're going to learn how we can grow closer to Christ by reading the Bible.

ACTIVITY 1: Continuous Change

Point: We need to grow closer to Jesus each day.

Supplies: You'll need a pitcher, lemonade mix (sugarless), sugar, dry ice, and a Bible.

Activity: Make a pitcher of lemonade without sugar and give a glass to each family member. Have them drink some of the lemonade and then share how it tasted.

Read aloud Acts 9:1-18. Then consider these questions:

- **Saul was bitter toward Christians. What does it mean to be bitter toward someone?** (To not like them; it means you are mean to them.)
- **How did Saul react to people who followed Jesus?** (He hated them; he tried to hurt them.)

Add sugar to each cup of lemonade and have children mix it

together and drink some more. Again, ask how the lemonade tastes.

Share: We had to add something to our lemonade to make it taste good—to take away the bitterness. God took away Saul's bitterness when Saul became a Christian. That's when he became Paul—a man who helped to grow Christian churches for the rest of his life. When we add God's love to our lives, the bitter turns to sweet.

Age Adjustments

Older Children and Teenagers may enjoy this activity more if done using popcorn. Have your children prepare plain popcorn (using an air popper or on the stove-top using unflavored oil) and taste it. Then have them add butter and taste it a second time; add salt and taste it a third time. While the "ooh" and "ahh" factor provided by the dry ice won't be a factor, older children might enjoy the food better and be able to distinguish the subtle differences between the plain, buttered, and buttered and salted popcorn. Talk with your children about the small ways they can grow closer to God. Help them see that even little things like more prayer time can add greatly to their walk with God.

Use gloves or tongs to add dry ice to the drinks. NOTE: Do not touch the dry ice with your fingers or allow children to touch it. As the dry ice is added, the lemonade will start to give off a vapor. Let it set a couple minutes, then have family members taste the drink again. This time, it will taste like a carbonated drink.

Consider these questions:

- **How does this compare to the lemonade with the sugar?** (It's better; it's more fun.)
- **How is the way we added more things to make the lemonade better like the way we need to add to our lives to make them better?** (We need to grow in God's love; we need to look for more ways to make life taste better.)

Share: Once we choose to follow Christ, we begin a journey of faith. We need to look for ways to grow closer to God every day. Just as we added sugar, then carbonation to our lemonade, we need to add friends, prayer, studying the Bible, going to church, and other things to grow more like Jesus.

ACTIVITY 2: The Raisin for Bible Study

Point: We need to feed on God's Word to grow in Christ.

Supplies: You'll need raisins, a clear drinking glass, a 2-liter bottle of a clear soft drink (such as 7-Up or Sprite), and a Bible.

Activity: Have volunteers look up and **read** aloud the following verses: Psalm 119:105; 2 Chronicles 34:31; Acts 17:11; and

James 1:22-25. Then consider the following questions:

- **Why is it important to spend time reading God's Word?** (To grow closer to Him; because God wants us to.)
- **What are some of the things we learn about reading the Bible from these passages?** (The Bible helps us know what to do; we should be excited to read the Bible; the Bible can help us in lots of ways.)

Pour the clear soft drink into a clear glass. Then **share: I'm going to drop these raisins in the glass. What do you think will happen to the raisins?** (They will float; they will sink.)

Pour the raisins into the glass and watch as they sink, gradually rise to the top, and sink back down to the bottom of the glass.

 Ask:

- **How are we like these "dancing raisins"?** (We have good days and bad days; sometimes we're up and sometimes we're down.)

Share: When we don't feed on God's Word every day, we sink down and our spiritual life suffers. But God wants our hearts and minds every day! He wants us to be floating at the top of our relationship with Him always.

Close with a time of prayer, asking God to help each family member look to the Bible for guidance each day—so they won't have as many "sinking" days in the future.

WRAP-UP

Gather everyone in a circle and have family members take turns answering this question: **What's one thing you've learned about God today?**

Next, tell kids you've got a new "Life Slogan" you'd like to share with them.

Life Slogan: Today's Life Slogan is this: "The Word and Spirit are our link; without them we would surely sink." Have family members repeat the slogan two or three times to help them learn it. Then encourage them to practice saying it during the week so they can talk about it at your next family night session.

Close in Prayer: Allow time for each family member to share prayer concerns and answers to prayer. Then close your time together with prayer for each concern. Thank God for listening to and caring about us.

Remember to record your prayer requests so you can refer to them in the future as you see God answering them.

@ 7: Word Power

Exploring how our words can be powerful weapons or godly tools

Scripture
- James 3:5-8
- Psalm 66:1
- Psalm 81:1
- Psalm 95:1
- Psalm 98:4
- Psalm 100:1

ACTIVITY OVERVIEW		
Activity	**Summary**	**Pre-Session Prep**
Activity 1: Taming the Tongue	Discover the most powerful weapon in the world.	You'll need a squirt gun, a pie pan, Pop Rocks candy, and a Bible.
Activity 2: Joyful Noise	Praise God with home-made instruments.	You'll need plastic straws, scissors, and a Bible.

Main Points:

—Our tongue is powerful and should be used to glorify God.
—It is important to spend time praising God.

LIFE SLOGAN: "Use your tongue as you should; for God's glory and good."

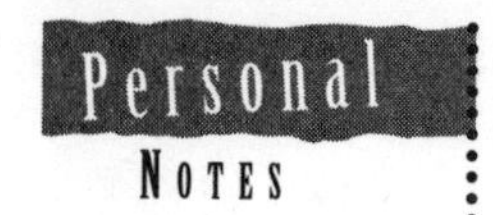

Make it your own
In the space provided below, outline the flow and add any additional ideas to guide you through the process of conducting this family night.

Prayer & Praise Items
In the space provided below, list any items you wish to pray about or give praise for during this family night session.

Journal
In the space provided below, capture a record of any fun or meaningful things which happened during this family night session.

Session Tip

We intentionally have provided more material than we would expect to be used in a single "Family Night" session. You know your family's unique interests and life circumstances best, so feel free to adapt this lesson to meet your family members' needs. Remember, short and simple is better than long and comprehensive.

WARM-UP

Open with Prayer: Begin by having a family member pray, asking God to help everyone in the family understand more about Him through this time. After prayer, review your last lesson by asking these questions:

- **What did we learn about in our last lesson?**
- **What was the Life Slogan?**
- **Have your actions changed because of what we learned? If so, how?** Encourage family members to give specific examples of how they've applied learning from the past week.

Share: Today we're going to learn about the power of our words—and how we can use our words to glorify God.

ACTIVITY 1: Taming the Tongue

Point: Our tongue is powerful and should be used to glorify God.

Supplies: You'll need a squirt gun, a pie pan, Pop Rocks candy, and a Bible.

Activity: Set the squirt gun on a table next to a pie pan in which you've spread some Pop Rocks candy.

Ask:

- **What is the most powerful weapon in the world?** (A nuclear bomb; a bazooka; a big gun.)

Have someone squirt the Pop Rocks using the water pistol. Then consider the following questions:

- **How is the way this gun is making the candies pop like the way weapons hurt people?** (Guns shoot people; there are loud noises when guns go off; we can see the effects of the guns.)

Have each person put a small amount of Pop Rocks onto their tongues. Enjoy together the strange feeling of the carbonated candies popping in your mouths. Then consider the following questions:

- **What is different about the way the candies popped in your mouth, compared to how they popped when we shot them with the water gun?** (I can feel these; they seem to move more; I can hear the popping more.)

Age Adjustments

OLDER CHILDREN AND TEENAGERS may have lots of friends who use crude or unkind language. Use this opportunity to talk with them about how they feel when someone cuts another person down or offers up a swear word every three words. Help your older children discover that their decision to use positive words can actually help friends see that bad language isn't necessary.

Share: Guns, bombs, and missiles are all very powerful weapons. And just as we could see the effects of our water gun on the candies, we can see the powerfully negative effects of these weapons on other people. But there is a more powerful weapon. Do you know what it is? Your tongue! Just as the effects of the Pop Rocks seemed greater when they were in your mouth, the power of your mouth—and your words—can be greater than any bullet or bomb.

Read James 3:5-8. **Share: Our tongue is powerful—more powerful than any weapon. But while weapons can only harm people and things, our tongue can actually help people.**

Have family members share some of the destructive and constructive things people can do with words. Then **share: Sometimes we say things that hurt other people. With the Holy Spirit's help, we can control our words and say only things that help people. Before we say something, we should first ask three questions: Is it true? Is it positive? Is it necessary?**

Have someone use their tongue constructively by asking God to help each family member make good choices when talking.

ACTIVITY 2: A Joyful Noise

Point: It is important to spend time praising God.

Supplies: You'll need plastic straws, scissors, and a Bible.

Activity: Read aloud Psalms 66:1; 81:1; 95:1; 98:4; and 100:1. Then ask children to identify the four words that are common to all of the Bible verses. The answer, of course, is "Make a joyful noise."

Consider these questions:

- **What does it mean to make a joyful noise unto the Lord?** (Sing; praise God with music; make a loud noise that tells God you love Him.)
- **Why does the Bible encourage us to make a joyful noise?** (Because God loves us; because we love God; because God saved us.)

Share: We can praise God for all the wonderful things He has done for our family.

Have family members call out things they'd like to praise God for (such as having a place to live; good food; friends; the Bible; that He is always near). Then explain that you're going to have some fun and make a joyful noise together.

Give each family member a straw and scissors (you'll want to help younger children with this activity). Have them flatten one end of the straw and cut a v-shape into that end.

Then have family members place their mouths over the flattened ends (just barely covering the "v") and blow. They will make a "joyful" sound together! You may wish to experiment by cutting various lengths of straws to make different sounds, or cutting holes in the top for "fingering" notes as if playing a recorder or song flute. Have family members play songs for others to guess.

End the activity by having each person say one thing he or she praises God for, then close with a rousing straw-orchestra rendition of a favorite praise chorus.

WRAP-UP

Gather everyone in a circle and have family members take turns answering this question: **What's one thing you've learned about God today?**

Next, tell kids you've got a new "Life Slogan" you'd like to share with them.

Life Slogan: Today's Life Slogan is this: "Use your tongue as you should; for God's glory and good." Have family members repeat the slogan two or three times to help them learn it. Then encourage them to practice saying it during the week so they can talk about it at your next family night session.

Close in Prayer: Allow time for each family member to share prayer concerns and answers to prayer. Then close your time together with prayer for each concern. Thank God for listening to and caring about us.

Remember to record your prayer requests so you can refer to them in the future as you see God answering them.

8: Discernment

Exploring how to test what the world offers, and follow God

Scripture
- Romans 12:2
- 1 John 4:1

ACTIVITY OVERVIEW		
Activity	**Summary**	**Pre-Session Prep**
Activity 1: Don't Be a Pea	Watch how peas fall out of a glass and learn to follow God, not the world.	You'll need a regular glass, dried peas, a wine glass (or other similar glass), a pie tin, water, and a Bible.
Activity 2: Test Everything	Learn how to discern truth from a lie.	You'll need a candle, an apple, an almond, and a Bible.

Main Points:

—Our actions should mirror God, not the world.

—Test what the world offers for consistency with Jesus' teachings.

LIFE SLOGAN: "Put the world's ways to the test; follow God's Word, it's always best."

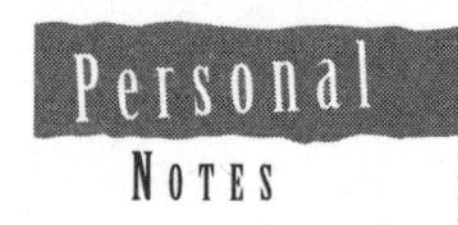

Make it your own

In the space provided below, outline the flow and add any additional ideas to guide you through the process of conducting this family night.

Prayer & Praise Items

In the space provided below, list any items you wish to pray about or give praise for during this family night session.

Journal

In the space provided below, capture a record of any fun or meaningful things which happened during this family night session.

Session Tip

We intentionally have provided more material than we would expect to be used in a single "Family Night" session. You know your family's unique interests and life circumstances best, so feel free to adapt this lesson to meet your family members' needs. Remember, short and simple is better than long and comprehensive.

WARM-UP

Open with Prayer: Begin by having a family member pray, asking God to help everyone in the family understand more about Him through this time. After prayer, review your last lesson by asking these questions:

- **What did we learn about in our last lesson?**
- **What was the Life Slogan?**
- **Have your actions changed because of what we learned? If so, how?** Encourage family members to give specific examples of how they've applied learning from the past week.

Share: Today we're going to learn why it's important to test things to determine the truth and why we must follow God and not the world.

ACTIVITY 1: Don't Be a Pea

Point: Our actions should mirror God, not the world.

Supplies: You'll need a regular glass, dried peas, a wine glass (or other similar glass), a pie tin, water, and a Bible.

Activity: Use this family night activity on a day when the family will be home most of the day. It's great for those rainy weekend days when outdoor activities aren't an option.

During the morning, **read** aloud Romans 12:2. Ask family members to share how they think the passage applies to them. Explain that being a Christian means choosing to live differently from those who do not know Christ.

Have volunteers share examples of how Christians live their lives differently than others. Then set the glass on the table; set the pie tin on top of the glass; set the wine glass on top of the pie tin; and fill the wine glass to overflowing with the peas. Then pour water up to the brim of the wine glass.

Ask family members to predict what will happen during the day to this strange construction. Leave the contraption set up in a prominent place in the house and continue with daily activities. Encourage family members to inspect the peas throughout the day to see what's happening to them. It may take several hours, but eventually, the heap of peas will become higher and then the peas will begin to fall (this could last for hours!). When the peas have begun to fall, call everyone together to discuss the following questions:

- **What are the peas doing?** (They're all falling out of the glass; they're playing "follow the leader.")
- **How are we like the peas?** (When one person does something, lots of people follow; we do what others are doing, not what God wants us to do.)
- **What does our Bible verse tell us about this kind of behavior?** (We need to be different from the world; our actions should not be like others'.)

Enjoy the cascade of peas awhile longer, then encourage family members to be confident and bold in following Christ—not following those who would lead them astray.

Age Adjustments

YOUNGER CHILDREN will enjoy this activity, but may not have the patience to wait for the peas to fall. Consider setting up a row of dominoes to tip over as a more immediate example of how people tend to follow the leader in life. Use this as a springboard into discussion about who they choose to follow: God or others. You may be able to apply this principle to a specific scenario they've recently faced. For example, if a child recently got in trouble for writing on the wall because he or she saw someone else do it, you could talk about how easy it is to follow others—but how it's not always what God wants.

ACTIVITY 2: Test Everything

Point: Test what the world offers for consistency with Jesus' teachings.

Supplies: You'll need a candle, an apple, an almond, and a Bible.

Activity: Place the candle and the apple next to each other and ask family members to describe what they see. Then explain that you see two candles. Have a volunteer look up the definition of a candle in the dictionary. It will read something like this: "A solid, usually cylindrical mass, with an embedded wick that is burned to produce light."

Share: I still see two candles. Let me show you what I mean.

Cut a cylindrical shape out of the apple, then embed the almond into the top of the shape. Explain that the apple now meets the general definition of a candle. **Share: As you grow in your faith, you will come across people who tell you they've got a message from God you must follow. But how do you know that what they're saying is true? How do you know if they really are telling you something from God? The same way you can tell if I'm telling you the truth about the apple: you test what they say.**

Light the almond to show family members that your apple can indeed be a candle. Then **read** 1 John 4:1 and ask: **How do we test what people say about God?**

Share: We need to test everything we hear from friends, radio programs, television, and even from our own church, to see if it is true. We test things by seeing if they match what is said in the Bible; by looking at the lifestyle of the person who tells us something; by seeing what other good things the person has done; and by learning about other beliefs that person has about God. If all these things hold true, with the guidance of the Holy Spirit, we can learn what is a truth and what is not. We must test everything to make sure we're following the truth.

WRAP-UP

Gather everyone in a circle and have family members take turns answering this question: **What's one thing you've learned about God today?**

Next, tell kids you've got a new "Life Slogan" you'd like to share with them.

Life Slogan: Today's Life Slogan is this: "Put the world's ways to the test; follow God's Word, it's always best." Have family members repeat the slogan two or three times to help them learn it. Then encourage them to practice saying it during the week so they can talk about it at your next family night session.

Close in Prayer: Allow time for each family member to share prayer concerns and answers to prayer. Then close your time together with prayer for each concern. Thank God for listening to and caring about us.

Remember to record your prayer requests so you can refer to them in the future as you see God answering them.

9: The Bondage of Sin

Exploring how we get tied up in sin

Scripture
- James 1:12-15
- Matthew 25:31-46

ACTIVITY OVERVIEW		
Activity	**Summary**	**Pre-Session Prep**
Activity 1: Breaking the Pattern of Sin	Learn how a continued life of sin is hard to break.	You'll need a supply of paper, a pan, water, and a Bible.
Activity 2: The Sheep and the Goats	Discover how to separate salt and pepper and learn about God's judgment.	You'll need coarse salt, ground pepper, a plastic spoon, a piece of wool cloth, and a Bible.

Main Points:

—It's not easy to break a pattern of sin.
—God will separate those who love Him from those who don't.

LIFE SLOGAN: "If we want to be a sheep; to the needs of others, we must leap."

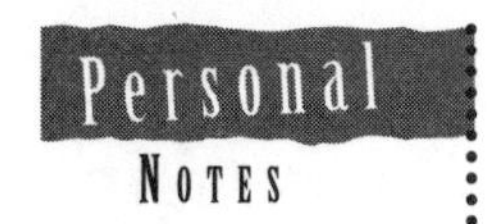

Make it your own

In the space provided below, outline the flow and add any additional ideas to guide you through the process of conducting this family night.

Prayer & Praise Items

In the space provided below, list any items you wish to pray about or give praise for during this family night session.

Journal

In the space provided below, capture a record of any fun or meaningful things which happened during this family night session.

Session Tip

We intentionally have provided more material than we would expect to be used in a single "Family Night" session. You know your family's unique interests and life circumstances best, so feel free to adapt this lesson to meet your family members' needs. Remember, short and simple is better than long and comprehensive.

WARM-UP

Open with Prayer: Begin by having a family member pray, asking God to help everyone in the family understand more about Him through this time. After prayer, review your last lesson by asking these questions:

- **What did we learn about in our last lesson?**
- **What was the Life Slogan?**
- **Have your actions changed because of what we learned? If so, how?** Encourage family members to give specific examples of how they've applied learning from the past week.

Share: Today we're going to learn how sin can tie us up and how God will separate those who love Him from those who don't.

ACTIVITY 1: Breaking the Pattern of Sin

Point: It's not easy to break a pattern of sin.

Supplies: You'll need a supply of paper, a pan, water, and a Bible.

Activity: Give each family member a sheet of paper and ask if they think they can tear that paper with their bare hands. After they answer, have everyone attempt to tear the paper. Then give them each a pile of 10 or more sheets of paper and have them attempt to tear those papers all at once. Repeat this action, adding paper to the piles until family members can no longer tear through them.

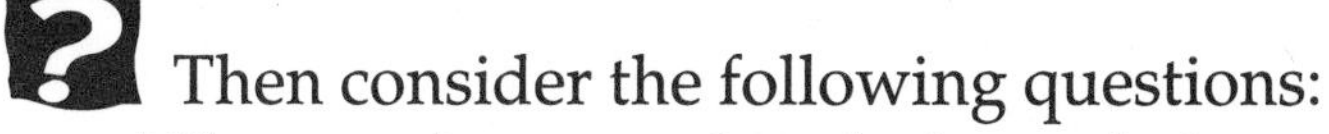

Then consider the following questions:

- **Why was it easy to break through the paper the first few times?** (Because it was thin; there wasn't much paper at first.)

- **What happened as we added more paper?** (It got harder to tear; I couldn't break it.)

Take the largest bunch of paper, write the word "our sins" on the top page, and set the pages in a pan of water. You'll come back to this in a few moments.

Age Adjustments

YOUNGER CHILDREN will be amazed by a simple variation on this activity. Rather than have them tear papers, use thread to tie up their wrists, beginning with one strand, then adding more until they can't pull their wrists apart. Use this activity to illustrate how it gets more and more difficult to break a pattern of sin as time goes on. Then cut the thread with scissors to illustrate how the pattern can be broken if we rely on the Holy Spirit and the advice from Christian friends and family members.

Share: Imagine that each of those sheets of paper is a little sin. When we have little sin in our lives, it's easy to break through and do the right thing. But if we continue to add sin, it becomes more and more difficult to break through—just as it was more difficult to tear through the large pile of paper. On their own, each little sin doesn't seem like too much to handle, but when we continue in a pattern of sin, it becomes more difficult to break away.

Have family members share ways they can avoid a pattern of sin. Then have them offer suggestions for getting out of a pattern of sin. For example, someone might suggest having a friend help, reading the Bible, or consulting with a pastor.

Read aloud James 1:12-15. **Share: It is a good thing to stay away from a life of sin. All of us sin. But when we do, we can ask forgiveness.**

Take the now-soaked papers and show family members how easy it is to tear them. Then **share: On our own, it's not easy to break a pattern of sin, but when we're soaked in the power of the Holy Spirit, and bathed in the love of friends and family, we can break through.**

ACTIVITY 2: The Sheep and the Goats

Point: God will separate those who love Him from those who don't.

Supplies: You'll need coarse salt, ground pepper, a plastic spoon, a piece of wool cloth, and a Bible.

Activity: Pour a bunch of coarse salt and ground pepper onto the table and mix them together. Then invite family members to see if they can separate the pepper from the salt. Encourage them to try, even if they think they can't. After a moment or two, interrupt by asking:

- **What is difficult about this activity?** (The salt and pepper are mixed together too well; it's hard to pick up the small pieces of salt.)

Read the story of the sheep and the goats in Matthew 25:31-46. Then **share: In this story, Jesus used the idea of sheep and goats because these animals were similar. They would often graze in the same fields, but they would be separated when it was time for shearing.**

Explain that the sheep and the goats, like the salt and pepper, represent people. Ask family members to imagine that the goats and the pepper are people who don't love God and that the sheep and the salt are people who do.

Then ask:

- **What did Jesus say to the sheep (or salt)?** (You get eternal life; I was sick and you took care of Me.)
- **What excuse did the goats (or pepper) have when God separated them?** (We didn't see when God was hungry; we never saw when God was in prison.)

Share: This passage tells us that when we serve people here on earth, we're also serving Jesus. When we serve Him, we get the prize of eternal life. Now let's go back and see if we can separate the salt and pepper again. As we do, let's remember why it's important to be the salt or the sheep—to serve God by serving others.

Take the plastic spoon and rub it with the wool cloth, then hold it over the mixture. The pepper will jump up onto the spoon and remain there. Throw away the pepper and use the salt that evening to season your food as a reminder of the importance of being like the salt or sheep.

WRAP-UP

Gather everyone in a circle and have family members take turns answering this question: **What's one thing you've learned about God today?**

Next, tell kids you've got a new "Life Slogan" you'd like to share with them.

Life Slogan: Today's Life Slogan is this: "If we want to be a sheep; to the needs of others, we must leap." Have family members repeat the slogan two or three times to help them learn it. Then encourage them to practice saying it during the week so they can talk about it at your next family night session.

Close in Prayer: Allow time for each family member to share prayer concerns and answers to prayer. Then close your time together with prayer for each concern. Thank God for listening to and caring about us.

Remember to record your prayer requests so you can refer to them in the future as you see God answering them.

10: A Passion for God

Exploring how to have a passionate love for God

Scripture
- Revelation 3:16
- Luke 12:31
- Mark 1:35
- Luke 4:16
- Mark 13:31

ACTIVITY OVERVIEW		
Activity	**Summary**	**Pre-Session Prep**
Activity 1: Neither Hot nor Cold	Learn how drinks taste better when they're hot or cold, and how God wants us to choose to love Him.	You'll need a pan of hot water, a pan of cold water, a pan of lukewarm water, a variety of hot and cold drinks, and a Bible.
Activity 2: Which One First?	Place objects in a jar and learn about priorities.	You'll need a wide-mouth glass jar, large rocks (golf-ball size), sand, water, a permanent marker, and a Bible.

Main Points:

—God wants a passionate relationship with us.

—Putting God first builds a solid relationship.

LIFE SLOGAN: "Keeping Him first; drives our thirst."

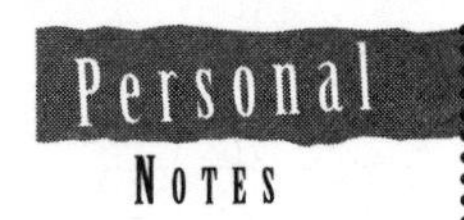

Make it your own

In the space provided below, outline the flow and add any additional ideas to guide you through the process of conducting this family night.

Prayer & Praise Items

In the space provided below, list any items you wish to pray about or give praise for during this family night session.

Journal

In the space provided below, capture a record of any fun or meaningful things which happened during this family night session.

Session Tip

We intentionally have provided more material than we would expect to be used in a single "Family Night" session. You know your family's unique interests and life circumstances best, so feel free to adapt this lesson to meet your family members' needs. Remember, short and simple is better than long and comprehensive.

WARM-UP

Open with Prayer: Begin by having a family member pray, asking God to help everyone in the family understand more about Him through this time. After prayer, review your last lesson by asking these questions:

- **What did we learn about in our last lesson?**
- **What was the Life Slogan?**
- **Have your actions changed because of what we learned? If so, how?** Encourage family members to give specific examples of how they've applied learning from the past week.

Share: Today we're going to learn that God wants us to be passionate about our relationship with Him, and that we should seek to put God first in our lives.

ACTIVITY 1: Neither Hot nor Cold

Point: God wants a passionate relationship with us.

Supplies: You'll need a pan of hot water, a pan of cold water, a pan of lukewarm water, a variety of hot and cold drinks—all to be served at room temperature, and a Bible.

Activity: Serve family members each a lukewarm drink that would taste much better if it were cold or hot (such as lemonade, "hot" chocolate, tea, and so forth). As family members are tasting their drinks, consider these questions:

- **What would have made these drinks taste better?** (If they were hot or cold, not room temperature.)
- **What are other things that taste better when they're hot or cold, but not in-between?** (I like hot french fries; warm ice cream wouldn't taste very good; I prefer hot pizza.) If possible, serve some of these foods at room temperature, to help

family members understand the term "lukewarm" as described in the Scripture passage.

Read aloud Revelation 3:16 and ask:

- **What does the author mean when he tells the church of Laodicea that they are "lukewarm"?** (They're not "hot" or "cold" toward God; they don't care about anything.)
- **In what ways are we lukewarm for Christ?** (We don't always live like Jesus wants us to; we don't do anything to grow in faith.)

Place three pans of water on the table: one with hot water (make sure the hot water isn't scalding hot—just plenty warm to feel the temperature); one with room-temperature water; and one with ice-cold water. Have family members each take a turn with this activity:

1. Place one hand each in the hot and cold water and leave them there for as long as possible (up to 2 minutes).
2. Remove your hands and immediately place them both in the pan with room-temperature water.
3. Close your eyes.
4. Answer the question: What is the temperature of the water for each hand? (The hand that was in the ice water will feel hot and the hand that was in the hot water will feel cold.)

Age Adjustments

YOUNGER CHILDREN may be afraid to put their hands in hot water—as they've likely been warned about the dangers of scalding. Make sure the water temperature is appropriate for younger hands, and assure children that this activity is safe.

Have everyone take their hands out of the lukewarm water for a few minutes. Then **share: The church at Laodicea was neither hot nor cold in their faith. Place your hands back in the room-temperature water and notice how the water doesn't feel hot or cold anymore. That's how this church seemed to feel about God. Sometimes we're this way too, we aren't doing anything against God, but we aren't doing anything to grow closer to Him, either.**

Have family members brainstorm ways to prevent becoming lukewarm or cold in their relationship with God.

ACTIVITY 2: Which One First?

Point: Putting God first builds a solid relationship.

Supplies: You'll need a wide-mouth glass jar, large rocks (golf-ball size), sand, water, a permanent marker, and a Bible.

Activity: Prior to this activity, test the amounts of rocks, sand, and water that will fit into the jar. First, place the large rocks in the jar. Then pour in the sand and shake it down to fill in the cracks. Finally, pour in enough water to fill up the jar. You'll need a new supply of sand and water to redo this activity with your family, but you should be able to wash off the rocks and the jar to be reused.

Ask your family members to figure tell you how all the items (rocks, pebbles, sand, and water) might fit into the jar. They may not believe it's possible or may suggest placing the pebbles, sand, or water in first. Listen to their ideas, then ask:

- **What are activities that occupy your time?** (School; eating; sleeping; playing with friends.) Pour the sand into the jar for each of the items listed until all the sand is in the jar.

 Then ask:

- **What are the things God would have us do with our time?**

 (**Read** the following verses to help with brainstorming: Mark 1:35; Luke 4:16; Mark 13:31)

Using the marker, label the larger rocks with these activities we should spend time on, such as Bible study; prayer; Sunday School; sharing about faith; and other topics. Attempt to place these in the jar (they won't fit well with the pebbles or sand already in place).

Share: It looks like we've run out of room in the jar . . . just like we run out of room in our lives by putting other things ahead of prayer, Bible study, and things that bring us closer to God.

Remove the rocks and sand. Place the rocks in first this time, and then pour in the sand, shaking it so it will settle around the rocks. Then pour in the water until the jar is full.

 Consider these questions:

- **How does this compare to the first attempt we had to put things in the jar?** (It worked this time; we put the big things in first.)
- **Why is it important to put the most important things in our lives first?** (Otherwise, they might not fit; the rest can flow around those things.)

Read aloud Luke 12:31 and **share: When we seek God first, the rest of our lives fall into place better. Putting God last means we probably won't have any time for Him at all—just as we didn't have space for the larger rocks when we put in the sand first. We must learn to thirst for God first!**

WRAP-UP

Gather everyone in a circle and have family members take turns answering this question: **What's one thing you've learned about God today?**

Next, tell kids you've got a new "Life Slogan" you'd like to share with them.

Life Slogan: Today's Life Slogan is this: "Keeping Him first; drives our thirst." Have family members repeat the slogan two or three times to help them learn it. Then encourage them to practice saying it during the week so they can talk about it at your next family night session.

Close in Prayer: Allow time for each family member to share prayer concerns and answers to prayer. Then close your time together with prayer for each concern. Thank God for listening to and caring about us.

Remember to record your prayer requests so you can refer to them in the future as you see God answering them.

11: Resting on the Promises

Exploring how God keeps His promises

Scripture
- Genesis 6–9:16
- Matthew 28:20

ACTIVITY OVERVIEW

Activity	Summary	Pre-Session Prep
Activity 1: Promises, Promises	Make a rainbow and remember how God kept His promise to Noah.	You'll need a plastic coffee can lid, a flashlight, bubble solution (see text), a straw, and a Bible.
Activity 2: Always with Us	Make a Möbius strip and learn that God's love lasts forever	You'll need a long sheet of paper, a pencil, scissors, tape or glue, and a Bible.

Main Points:

—God keeps His promises.

—God will never leave nor forsake us.

LIFE SLOGAN: "We need to believe; that God will never leave."

Make it your own

In the space provided below, outline the flow and add any additional ideas to guide you through the process of conducting this family night.

Prayer & Praise Items

In the space provided below, list any items you wish to pray about or give praise for during this family night session.

Journal

In the space provided below, capture a record of any fun or meaningful things which happened during this family night session.

Session Tip

We intentionally have provided more material than we would expect to be used in a single "Family Night" session. You know your family's unique interests and life circumstances best, so feel free to adapt this lesson to meet your family members' needs. Remember, short and simple is better than long and comprehensive.

WARM-UP

Open with Prayer: Begin by having a family member pray, asking God to help everyone in the family understand more about Him through this time. After prayer, review your last lesson by asking these questions:

- **What did we learn about in our last lesson?**
- **What was the Life Slogan?**
- **Have your actions changed because of what we learned? If so, how?** Encourage family members to give specific examples of how they've applied learning from the past week.

Share: Today we're going to discover that God keeps His promises and that He'll always be with us.

ACTIVITY 1: Promises, Promises

Point: God keeps His promises.

Supplies: You'll need a plastic coffee can lid, a flashlight, bubble solution (see text), a straw, and a Bible.

Activity: Read or summarize the account of Noah and the flood in Genesis 6–9:16. Emphasize the focus of Genesis 9:12-16. Then consider these questions:

- **Why did God flood the world?** (Because nobody loved God; people had forgotten God; it was full of sin.)
- **Why did God save Noah and his family?** (Because they loved God; they were the only people who followed God.)
- **What does it mean to be "godly"?** (To be like God; to follow God.)

Have family members share ideas on what it means to be godly. Then talk about how Noah kept working even when the other people were laughing at him and making fun of him. Ask your children when they've felt like Noah's family. Then share that being godly means standing for what you believe even when others don't agree or make fun of you.

Tell family members that God made a promise to never flood the earth again and that the rainbow would be a reminder of that promise. Then explain that you're going to make a rainbow as a reminder to your family that we should pursue godliness, just like Noah's family.

Set or hold the flashlight upright on a table. Balance the plastic coffee lid on the flashlight. Pour a spoonful of bubble liquid into the lid (see the recipe in the margin). Wet the lid and the straw with the bubble liquid. Turn on the flashlight and turn off the room lights. Use the straw to gently blow a bubble on the lid. Pull the straw out of the bubble and watch. Soon the bubble will get thin on top and thick on the bottom. Keep watching and you will be able to see a rainbow of colors in the bubble.

Just for fun: Have a child get a finger really wet and carefully push it into the bubble . . . the bubble won't break! As you do this, remind family members that God won't break His promises, either.

End this activity by asking family members to remember two things every time they see a rainbow: that God keeps His promises, and that God wants us to be godly.

Recipe for Bubble Liquid

Ad 2/3 cup of liquid dish soap to a gallon of water. Add the soap last, so you don't end up with a bunch of suds. Then add 1 tablespoon of glycerine (available at your local pharmacy). This will help your bubbles last longer. Use a dish soap that is clear or transparent for best results (Ajax and Dawn work well).

Age Adjustments

YOUNGER CHILDREN will enjoy the bubble activity, but may also enjoy playing with light and a prism (check your local nature or educational toy stores). Show children how the prism shines a rainbow when placed in the light. Use a flashlight to shine a rainbow on each family member, taking time to pray for each person to be godly like Noah and thanking God that He keeps His promises.

ACTIVITY 2: Always with Us

Point: God will never leave nor forsake us.

Supplies: You'll need a long sheet of paper, a pencil, scissors, tape or glue, and a Bible.

Activity: Read Matthew 28:20. Ask family members to tell about some of God's promises. You may wish to refer to a children's Bible for this. Here are a few things family members may come up with:

- God will never flood the whole world again.
- God loves us.
- God is always near.
- Those who love God will be with Him in heaven.

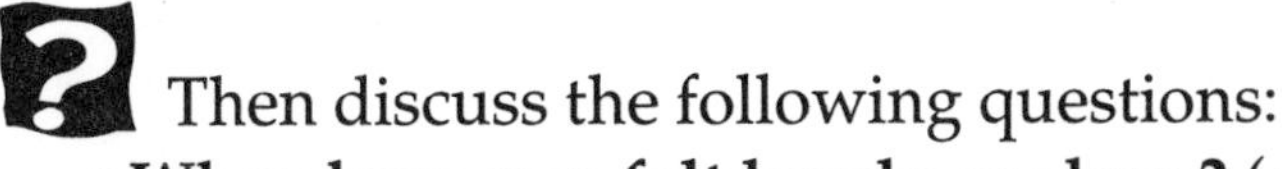

Then discuss the following questions:

- **When have you felt lonely or alone?** (Answers will vary.)
- **Who is with us, even when we feel alone?** (God.)

Make a Möbius strip for each family member by cutting a 2″ by 10″ strip of paper; giving one end a twist, and taping or gluing the ends together (see illustration).

Have children draw a line down the middle of their ring until it circles around and comes back to where they began. They may be amazed at this, wondering how one line can get on both sides of the paper. Then give them an even greater surprise. Ask what they think will happen if you cut the strip down the middle. Then help children cut their strips carefully, following the drawn line. Instead of two loops, you'll have one larger Möbius strip. You may repeat this activity again with the larger strip to show how it continues to grow and never ends.

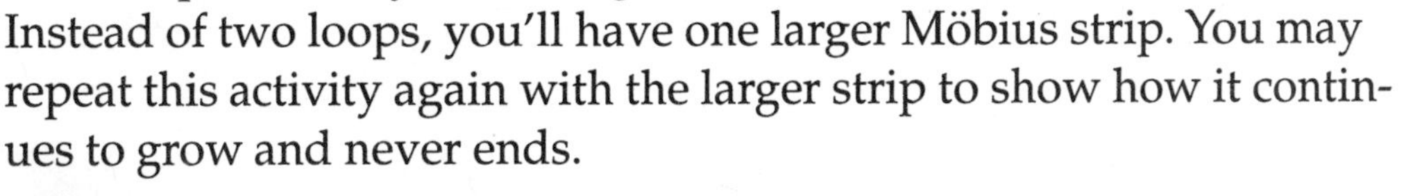

Then ask:

- **How is this strip like God?** (It never ends; nothing we do separates us from God; it goes on and on.)

Share: We may think that we've cut God out of our lives when we sin, but God is always near. He will keep His promise and walk beside us always. God never leaves. And like this strip, the more we get to know God, the "larger" He becomes.

WRAP-UP

Gather everyone in a circle and have family members take turns answering this question: **What's one thing you've learned about God today?**

Next, tell kids you've got a new "Life Slogan" you'd like to share with them.

Life Slogan: Today's Life Slogan is this: "We need to believe; that God will never leave." Have family members repeat the slogan two or three times to help them learn it. Then encourage them to practice saying it during the week so they can talk about it at your next family night session.

Close in Prayer: Allow time for each family member to share prayer concerns and answers to prayer. Then close your time together with prayer for each concern. Thank God for listening to and caring about us.

Remember to record your prayer requests so you can refer to them in the future as you see God answering them.

@ 12: Choosing Friends

Exploring how to choose friends who will help us grow in faith

Scripture
- 1 Samuel 16:7
- Galatians 2:6
- Ecclesiastes 4:9-12

ACTIVITY OVERVIEW		
Activity	Summary	Pre-Session Prep
Activity 1: Who Is My Friend?	Learn that it's what's on the "inside" that counts when choosing friends.	You'll need 4 cans of pop (2 diet, 2 regular), 1 large tub to put these in (a five-gallon bucket will do), duct tape, water, and a Bible.
Activity 2: Support System	Stack pennies on a piece of cardboard and learn the value of a supportive friend.	You'll need strips of cardboard cut out from a cereal box, books, 50 pennies, and a Bible.

Main Points:

—God looks at the heart.

—A good friend encourages us to do what Jesus would do.

LIFE SLOGAN: "God never lied; what's important is inside."

Make it your own
In the space provided below, outline the flow and add any additional ideas to guide you through the process of conducting this family night.

Prayer & Praise Items
In the space provided below, list any items you wish to pray about or give praise for during this family night session.

Journal
In the space provided below, capture a record of any fun or meaningful things which happened during this family night session.

Session Tip

We intentionally have provided more material than we would expect to be used in a single "Family Night" session. You know your family's unique interests and life circumstances best, so feel free to adapt this lesson to meet your family members' needs. Remember, short and simple is better than long and comprehensive.

WARM-UP

Open with Prayer: Begin by having a family member pray, asking God to help everyone in the family understand more about Him through this time. After prayer, review your last lesson by asking these questions:

- **What did we learn about in our last lesson?**
- **What was the Life Slogan?**
- **Have your actions changed because of what we learned? If so, how?** Encourage family members to give specific examples of how they've applied learning from the past week.

Share: Today we're going to learn about how to choose friends and how good friends can support us.

ACTIVITY 1: Who Is My Friend?

Point: God looks at the heart.

Supplies: You'll need 4 cans of pop (2 diet, 2 regular), 1 large tub to put these in (a five-gallon bucket will do), duct tape, water, and a Bible.

Activity: Prepare for this activity by taking the soft drink cans and wrapping them with duct tape so you can't see what kind of soft drinks they are.

Have children examine the soft drinks. Then fill the tub or bucket with water. Ask: **What do you think will happen when I place the cans in the water?** (They will sink; they will float.)

Put the cans in the water and watch as some float (the diet sodas) and some sink (the regular sodas).

Consider these questions:

- **How are the soft drinks like our friends?** (They look the same on the outside, but they're different on the inside; some people seem nice, but aren't.)
- **What do you think makes some of the cans float?** (More bubbles; more air in the can.) NOTE: The scientific reason is that the artificial sweetener used in the diet soda is lighter in weight than the sugar used in the regular soda.
- **What do you think makes some people good friends and others not-so-good friends?** (What they believe; how they act; what they think is important.)

Remove the duct tape and look at the soft drinks together. Then **read** aloud 1 Samuel 16:7 and Galatians 2:6.

Ask:

- **What does this passage tell us about people?** (It's what's on the inside that counts; our hearts are important to God.)
- **How can we make sure we "float" with God's love?** (Read the Bible; trust Jesus; go to church; pray.)

Open and enjoy the soft drinks together and **share: God tells us in the Bible that what's on the inside is what's most important to Him. When we look for friends, we need to look beyond their clothes and appearance to see what's inside.**

Ask family members to share ways they can discover what's inside someone. Then pray, asking God to help them make good choices when making friends.

Age Adjustments

Older Children and Teenagers have likely developed strong friendships by the time they've entered middle school. In an effort to feel wanted and liked, they may have chosen friends whose "insides" aren't what you'd prefer. Talk with older children and teenagers about their friends. Ask them to consider why they're friends and how they can boldly live out their faith among those friends. Use this activity as a discussion-starter on the importance of surrounding yourself with people who will help you grow spiritually—and the value of having non-Christian friends who can discover Christ through your witness and lifestyle.

ACTIVITY 2: Support System

Point: A good friend encourages us to do what Jesus would do.

Supplies: You'll need strips of cardboard cut out from a cereal box, books, 50 pennies, and a Bible.

Activity: Read Ecclesiastes 4:9-12. Cut 2" wide strips out of a cereal box, the height of the box. Place two stacks of books (about five inches high) on a table about seven inches apart and set a bunch of pennies next to the books. Send family members out of the room and have them return to attempt this activity one at a time.

Have every family member attempt to use one strip of cardboard and the two stacks of books to make a bridge. Then have them see how many pennies they can stack on the bridge before it falls. Repeat this activity so each person has a chance to stack the pennies. Keep track to see who is able to stack the most pennies.

After everyone has attempted this activity, gather together and show family members how easy it is to balance more pennies on the bridge. Take the second strip of cardboard and secure it as illustrated below (placing the edges of the cardboard at the bottom of the stacks of books, arching it upward until it touches the top piece of cardboard). Then have family members stack as many pennies as they can on the newly supported bridge.

 Ask:

- **What made this bridge sturdier than the others?** (The support of the other cardboard.)
- **How is this bridge like someone who has good, godly friends?** (They can support someone; good friends make you stronger.)

Share: By encouraging us to do what Jesus would do, standing by us, and helping us to seek God's will, good friends can support us and make us stronger.

Have family members share ways friends can be a positive influence (such as encouraging us to obey parents, doing things to show love to others, using positive words) and a negative influence (such as gossiping, hurting others, encouraging disobedience). Then show family members how bad friends can "let us down" by turning the support cardboard over into a "U" shape—the pennies will fall once again as the support is no longer there.

Close this activity by thanking God for giving us friends and asking Him for the wisdom to choose friends who will be supportive and positive influences.

WRAP-UP

Gather everyone in a circle and have family members take turns answering this question: **What's one thing you've learned about God today?**

Next, tell kids you've got a new "Life Slogan" you'd like to share with them.

Life Slogan: Today's Life Slogan is this: "God never lied; what's important is inside." Have family members repeat the slogan two or three times to help them learn it. Then encourage them to practice saying it during the week so they can talk about it at your next family night session.

Close in Prayer: Allow time for each family member to share prayer concerns and answers to prayer. Then close your time together with prayer for each concern. Thank God for listening to and caring about us.

Remember to record your prayer requests so you can refer to them in the future as you see God answering them.

How to Lead Your Child to Christ

SOME THINGS TO CONSIDER AHEAD OF TIME:

1. Realize that God is more concerned about your child's eternal destiny and happiness than you are. "The Lord is not slow in keeping His promise. . . . He is patient with you, not wanting anyone to perish, but everyone to come to repentance" (2 Peter 3:9).
2. Pray specifically beforehand that God will give you insights and wisdom in dealing with each child on his or her maturity level.
3. Don't use terms like "take Jesus into your heart," "dying and going to hell," and "accepting Christ as your personal Savior." Children are either too literal ("How does Jesus breathe in my heart?") or the words are too clichéd and trite for their understanding.
4. Deal with each child alone, and don't be in a hurry. Make sure he or she understands. Discuss. Take your time.

A FEW CAUTIONS:

1. When drawing children to Himself, Jesus said for others to "allow" them to come to Him (see Mark 10:14). Only with adults did He use the term "compel" (see Luke 14:23). Do not compel children.
2. Remember that unless the Holy Spirit is speaking to the child, there will be no genuine heart experience of regeneration. Parents, don't get caught up in the idea that Jesus will return the day before you were going to speak to your child about salvation and that it will be too late. Look at God's character—He *is* love! He is not dangling your child's soul over hell. Wait on God's timing.

   Pray with faith, believing. Be concerned, but don't push.

THE PLAN:

1. **God loves you.** Recite John 3:16 with your child's name in place of "the world."
2. **Show the child his or her need of a Savior.**
   a. Deal with sin carefully. There is one thing that cannot enter heaven—sin.

   b. Be sure your child knows what sin is. Ask him to name some (things common to children—lying, sassing, disobeying, etc.). Sin is doing or thinking anything wrong according to God's Word. It is breaking God's Law.

   c. Ask the question "Have you sinned?" If the answer is no, do not continue. Urge him to come and talk to you again when he does feel that he has sinned. Dismiss him. You may want to have prayer first, however, thanking God "for this young child who is willing to do what is right." Make it easy for him to talk to you again, but do not continue. Do not say, "Oh, yes, you have too sinned!" and then name some. With children, wait for God's conviction.

   d. If the answer is yes, continue. He may even give a personal illustration of some sin he has done recently or one that has bothered him.

   e. Tell him what God says about sin: We've all sinned ("There is no one righteous, not even one," Rom. 3:10). And because of that sin, we can't get to God ("For the wages of sin is death . . . " Rom. 6:23). So He had to come to us (". . . but the gift of God is eternal life in Christ Jesus our Lord," Rom. 6:23).

   f. Relate God's gift of salvation to Christmas gifts—we don't earn them or pay for them; we just accept them and are thankful for them.
3. **Bring the child to a definite decision.**
   a. Christ must be received if salvation is to be possessed.

   b. Remember, do not force a decision.

   c. Ask the child to pray out loud in her own words. Give her some things she could say if she seems unsure. Now be prepared for a blessing! (It is best to avoid having the child repeat a memorized prayer after you. Let her think, and make it personal.)*

d. After salvation has occurred, pray for her out loud. This is a good way to pronounce a blessing on her.

4. **Lead your child into assurance.**

   Show him that he will have to keep his relationship open with God through repentance and forgiveness (just like with his family or friends), but that God will always love him ("Never will I leave you; never will I forsake you," Heb. 13:5).

* If you wish to guide your child through the prayer, here is some suggested language.

> *"Dear God, I know that I am a sinner [have child name specific sins he or she acknowledged earlier, such as lying, stealing, disobeying, etc.]. I know that Jesus died on the cross to pay for all my sins. I ask You to forgive me of my sins. I believe that Jesus died for me and rose from the dead, and I accept Him as my Savior. Thank You for loving me. In Jesus' name. Amen."*

Cumulative Topical Index

AN INTRODUCTION TO FAMILY NIGHTS = IFN

BASIC CHRISTIAN BELIEFS = BCB

CHRISTIAN CHARACTER QUALITIES = CCQ

WISDOM LIFE SKILLS = WLS

MONEY MATTERS FOR KIDS = MMK

HOLIDAYS FAMILY NIGHT = HFN

BIBLE STORIES FOR PRESCHOOLERS (OLD TESTAMENT) = OTS

SIMPLE SCIENCE = SS

TOPIC	SCRIPTURE	WHAT YOU'LL NEED	WHERE TO FIND IT
Change Helps Us Grow and Mature	Rom. 8:28-39	Bible	WLS, p. 39
Change Is Good	1 Kings 17:8-16	Jar or box for holding change, colored paper, tape, markers, Bible	MMK, p. 27
Christ Is Who We Serve	Col. 3:23-24	Paper, scissors, pens	IFN, p. 50
Christians Should Be Joyful Each Day	James 3:22-23; Ps. 118:24	Small plastic bottle, cork to fit bottle opening, water, vinegar, paper towel, Bible	CCQ, p. 67
Commitment and Hard Work Are Needed to Finish Strong	Gen. 6:5-22	Jigsaw puzzle, Bible	CCQ, p. 83
The Consequence of Sin Is Death	Ps. 19:1-6	Dominoes	BCB, p. 57
Contentment Is the Secret to Happiness	Matt. 6:33	Package of candies, a Bible	MMK, p. 51
Creation	Gen. 1:1; Ps. 19:1-6; Rom. 1:20	Nature book or video, Bible	IFN, p. 17
David and Bathsheba	2 Sam. 11:1–12:14	Bible	BCB, p. 90
Description of Heaven	Rev. 21:3-4, 10-27	Bible, drawing supplies	BCB, p. 76
Difficulty Can Help Us Grow	Jer. 32:17; Luke 18:27	Bible, card game like Old Maid or Crazy Eights	CCQ, p. 33
Discipline and Training Make Us Stronger	Prov. 4:23	Narrow doorway, Bible	CCQ, p. 103
Do Not Give In to Those Around You	Matt. 14:6-12; Luke 23:13-25	One empty two-liter plastic bottle, eye-dropper, water, a Bible	SS, p. 21
Don't Be Yoked with Unbelievers	2 Cor. 16:17–17:1	Milk, food coloring	IFN, p. 105
Don't Give Respect Based on Material Wealth	Eph. 6:1-8; 1 Peter 2:13-17; Ps. 119:17; James 2:1-2; 1 Tim. 4:12	Large sheet of paper, tape, a pen, Bible	IFN, p. 64
Easter Was God's Plan for Jesus	John 3:16; Rom. 3:23; 6:23	Paper and pencils or pens, materials to make a large cross, and a Bible	HFN, p. 27

An Introduction to Family Nights = IFN

Basic Christian Beliefs = BCB

Christian Character Qualities = CCQ

Wisdom Life Skills = WLS

Money Matters for Kids = MMK

Holidays Family Night = HFN

Bible Stories for Preschoolers (Old Testament) = OTS

Simple Science = SS

Heritage Builders

AN INTRODUCTION TO FAMILY NIGHTS = IFN
BASIC CHRISTIAN BELIEFS = BCB
CHRISTIAN CHARACTER QUALITIES = CCQ
WISDOM LIFE SKILLS = WLS
MONEY MATTERS FOR KIDS = MMK
HOLIDAYS FAMILY NIGHT = HFN
BIBLE STORIES FOR PRESCHOOLERS (OLD TESTAMENT) = OTS
SIMPLE SCIENCE = SS

TOPIC	SCRIPTURE	WHAT YOU'LL NEED	WHERE TO FIND IT
Equality Does Not Mean Contentment	Matt. 20:1-16	Money or candy bars, tape recorder or radio, Bible	WLS, p. 21
Even if We're Not in the Majority, We May Be Right	2 Tim. 3:12-17	Piece of paper, pencil, water	CCQ, p. 95
Every Day Is a Gift from God	Prov. 16:9	Bible	CCQ, p. 69
Evil Hearts Say Evil Words	Prov. 15:2-8; Luke 6:45; Eph. 4:29	Bible, small mirror	IFN, p. 79
Family Members Ought to Be Loyal to Each Other	The Book of Ruth	Shoebox, two pieces of different colored felt, seven pipe cleaners (preferably of different colors)	OTS, p. 67
The Fruit of the Spirit	Gal. 5:22-23; Luke 3:8; Acts 26:20	Blindfold and Bible	BCB, p. 92
God Allows Testing to Help Us Mature	James 1:2-4	Bible	BCB, p. 44
God Became a Man So We Could Understand His Love	John 14:9-10	A pet of some kind, and a Bible	HFN, p. 85
God Can Clean Our Guilty Consciences	1 John 1:9	Small dish of bleach, dark piece of material, Bible	WLS, p. 95
God Can Do the Impossible	John 6:1-14	Bible, sturdy plank (6 or more inches wide and 6 to 8 feet long), a brick or similar object, snack of fish and crackers	CCQ, p. 31
God Can Give Us Strength		Musical instruments (or pots and pans with wooden spoons) and a snack	OTS, p. 52
God Can Guide Us Away from Satan's Traps	Ps. 119:9-11; Prov. 3:5-6	Ten or more inexpensive mousetraps, pencil, blindfold, Bible	WLS, p. 72
God Can Help Us Knock Sin Out of Our Lives	Ps. 32:1-5; 1 John 1:9	Heavy drinking glass, pie tin, small slips of paper, pencils, large raw egg, cardboard tube from a roll of toilet paper, broom, masking tape, Bible	WLS, p. 53
God Can Use Us in Unique Ways to Accomplish His Plans		Strings of cloth, clothespins or strong tape, "glow sticks" or small flashlights	OTS, p. 63

TOPIC	SCRIPTURE	WHAT YOU'LL NEED	WHERE TO FIND IT
God Cares for Us Even in Hard Times	Job 1–2; 42	Bible	WLS, p. 103
God Chose to Make Dads (or Moms) as a Picture of Himself	Gen. 1:26-27	Large sheets of paper, pencils, a bright light, a picture of your family, a Bible	HFN, p. 47
God Created the Heavens and the Earth	Gen. 1	Small tent or sheet and a rope, Christmas lights, two buckets (one with water), a coffee can with dirt, a tape recorder and cassette, and a flashlight	OTS, p. 17
God Created Us	Isa. 45:9, 64:8; Ps. 139:13	Bible and video of potter with clay	BCB, p. 43
God Created the World, Stars, Plants, Animals, and People	Gen. 1	Play dough or clay, safe shaping or cutting tools, a Bible	OTS, p. 19
God Doesn't Want Us to Worry	Matt. 6:25-34; Phil. 4:6-7; Ps. 55:22	Bible, paper, pencils	CCQ, p. 39
God Forgives Those Who Confess Their Sins	1 John 1:9	Sheets of paper, tape, Bible	BCB, p. 58
God Gave Jesus a Message for Us	John 1:14,18; 8:19; 12:49-50	Goldfish in water or bug in jar, water	BCB, p. 66
God Gives and God Can Take Away	Luke 12:13-21	Bible, timer with bell or buzzer, large bowl of small candies, smaller bowl for each child	CCQ, p. 15
God Gives Us the Skills We Need to Do What He Asks of Us		Materials to make a sling (cloth, shoestrings), plastic golf balls or marshmallows, stuffed animals	OTS, p. 73
God Is Holy	Ex. 3:1-6	Masking tape, baby powder or corn starch, broom, Bible	IFN, p. 31
God Is Invisible, Powerful, and Real	John 1:18, 4:24; Luke 24:36-39	Balloons, balls, refrigerator magnets, Bible	IFN, p. 15
God Is the Source of Our Strength	Jud. 16	Oversized sweatshirt, balloons, mop heads or other items to use as wigs, items to stack to make pillars, a Bible	OTS, p. 61
God Is Our Only Source of Strength	Isa. 40:29-31	Straws, fresh baking potatoes, a Bible	SS, p. 33

An Introduction to Family Nights = IFN

Basic Christian Beliefs = BCB

Christian Character Qualities = CCQ

Wisdom Life Skills = WLS

Money Matters for Kids = MMK

Holidays Family Night = HFN

Bible Stories for Preschoolers (Old Testament) = OTS

Simple Science = SS

Heritage Builders

An Introduction to Family Nights = IFN

Basic Christian Beliefs = BCB

Christian Character Qualities = CCQ

Wisdom Life Skills = WLS

Money Matters for Kids = MMK

Holidays Family Night = HFN

Bible Stories for Preschoolers (Old Testament) = OTS

Simple Science = SS

TOPIC	SCRIPTURE	WHAT YOU'LL NEED	WHERE TO FIND IT
God Is with Us	Ex. 25:10-22; Deut. 10:1-5; Josh. 3:14-17; 1 Sam. 3:3; 2 Sam. 6:12-15	A large cardboard box, two broom handles, a utility knife, strong tape, gold spray paint, and a Bible	OTS, p. 49
God Keeps His Promises	Gen. 6–9:16	Plastic coffee can lid, flashlight, bubble solution, straw, a Bible	SS, p. 75
God Keeps His Promises	Gen. 9:13, 15	Sheets of colored cellophane, cardboard, scissors, tape, a Bible, a lamp or large flashlight	OTS, p. 25
God Knew His Plans for Us	Jer. 29:11	Two puzzles and a Bible	BCB, p. 19
God Knew Moses Would Be Found by Pharaoh's Daughter	Ex. 2:1-10	A doll or stuffed animal, a basket, and a blanket	OTS, p. 43
God Knows All about Us	Ps. 139:2-4; Matt. 10:30	3x5 cards, a pen	BCB, p. 17
God Knows Everything	Isa. 40:13-14; Eph. 4:1-6	Bible	IFN, p. 15
God Knows the Plan for Our Lives	Rom. 8:28	Three different 25–50 piece jigsaw puzzles, Bible	WLS, p. 101
God Looks at the Heart	1 Sam. 16:7; Gal. 2:6	4 cans of pop (2 regular and 2 diet), 1 large tub, duct tape, water, a Bible	SS, p. 81
God Looks beyond the Mask and into Our Hearts		Costumes	HFN, p. 65
God Loves and Protects Us	Matt. 6:26-27	One or two raw eggs, a sink or bucket, a Bible	SS, p. 15
God Loves Us So Much, He Sent Jesus	John 3:16; Eph. 2:8-9	I.O.U. for each family member	IFN, p. 34
God Made Our Family Unique by Placing Each of Us in It		Different color paint for each family member, toothpicks or paintbrushes to dip into paint, white paper, Bible	BCB, p. 110
God Made Us		Building blocks, such as Tinkertoys, Legos, or K'nex	HFN, p. 15
God Made Us in His Image	Gen. 1:24-27	Play dough or clay and Bible	BCB, p. 24

TOPIC	SCRIPTURE	WHAT YOU'LL NEED	WHERE TO FIND IT
God Never Changes	Ecc. 3:1-8; Heb. 13:8	Paper, pencils, Bible	WLS, p. 37
God Owns Everything; He Gives Us Things to Manage		Large sheet of poster board or newsprint and colored markers	MMK, p. 17
God Provides a Way Out of Temptation	1 Cor. 10:12-13; James 1:13-14; 4:7; 1 John 2:15-17	Bible	IFN, p. 88
God Sees Who We Really Are—We Can Never Fool Him	1 Sam. 16:7	Construction paper, scissors, crayons or markers, a hat or bowl, and a Bible	HFN, p. 66
God Strengthens Us and Protects Us from Satan	2 Thes. 3:3; Ps. 18:2-3	Two un-inflated black balloons, water, a candle, matches, a Bible	SS, p. 16
God Teaches Us about Love through Others	1 Cor. 13	Colored paper, markers, crayons, scissors, tape or glue, and a Bible	HFN, p. 22
God Used Plagues to Tell Pharaoh to Let Moses and His People Go	Ex. 7–12	A clear glass, red food coloring, water, and a Bible	OTS, p. 44
God Uses Many Ways to Get Our Attention	Dan. 5	Large sheets of paper or poster board, tape, finger-paint, and a Bible	OTS, p. 79
God Wants Our Best Effort in All We Do	Col. 3:23-24	Children's blocks or a large supply of cardboard boxes	MMK, p. 63
God Wants a Passionate Relationship with Us	Rev. 3:16	Pans of hot, cold, and lukewarm water, hot and cold drinks	SS, p. 69
God Wants Us to Be Diligent in Our Work	Prov. 6:6-11; 1 Thes. 4:11-12	Video about ants or picture books or encyclopedia, Bible	CCQ, p. 55
God Wants Us to Get Closer to Him	James 4:8; 1 John 4:7-12	Hidden Bibles, clues to find them	BCB, p. 33
God Wants Us to Glorify Him	Ps. 24:1; Luke 12:13-21	Paper, pencils, Bible	WLS, p. 47
God Wants Us to Work and Be Helpful	2 Thes. 3:6-15	Several undone chores, Bible	CCQ, p. 53
God Will Never Leave Us or Forsake Us	Matt. 28:20	Long sheet of paper, pencil, scissors, tape or glue, a Bible	SS, p. 76
God Will Send the Holy Spirit	John 14:23-26; 1 Cor. 2:12	Flashlights, small treats, Bible	IFN, p. 39

An Introduction to Family Nights = IFN

Basic Christian Beliefs = BCB

Christian Character Qualities = CCQ

Wisdom Life Skills = WLS

Money Matters for Kids = MMK

Holidays Family Night = HFN

Bible Stories for Preschoolers (Old Testament) = OTS

Simple Science = SS

Heritage Builders

TOPIC	SCRIPTURE	WHAT YOU'LL NEED	WHERE TO FIND IT
God Will Separate Those Who Love Him from Those Who Don't	Matt. 25:31-46	Coarse salt, ground pepper, plastic spoon, wool cloth, a Bible	SS, p. 64
God's Covenant with Noah	Gen. 8:13-21; 9:8-17	Bible, paper, crayons or markers	BCB, p. 52
A Good Friend Encourages Us to Do What Jesus Would Do	Ecc. 4:9-12	Strips of cardboard, books, 50 pennies, a Bible	SS, p. 82
Guarding the Gate to Our Minds	Prov. 4:13; 2 Cor. 11:3; Phil. 4:8	Bible, poster board for each family member, old magazines, glue, scissors, markers	CCQ, p. 23
The Holy Spirit Helps Us	Eph. 1:17; John 14:15-17; Acts 1:1-11; Eph. 3:16-17; Rom. 8:26-27; 1 Cor. 2:11-16	Bible	BCB, p. 99
The Holy Spirit Helps Us to Be a Light in the Dark World	Matt. 5:14-16; 1 Tim. 2:1-4	Wintergreen or Cryst-O-Mint LifeSavers, a Bible	SS, p. 40
Honesty Means Being Sure We Tell the Truth and Are Fair	Prov. 10:9; 11:3; 12:5; 14:2; 28:13	A bunch of coins and a Bible	MMK, p. 58
Honor the Holy Spirit, Don't Block Him	1 John 4:4; 1 Cor. 6:19-20	Bible, blow-dryer or vacuum cleaner with exit hose, a Ping-Pong ball	CCQ, p. 47
Honor Your Parents	Ex. 20:12	Paper, pencil, treats, umbrella, soft objects, masking tape, pen, Bible	IFN, p. 55
How Big Is an Ark?		Large open area, buckets of water, cans of animal food, bags of dog food, and four flags	OTS, p. 24
If We Confess Our Sins, Jesus Will Forgive Us	Heb. 12:1;1 John 1:9	Magic slate, candies, paper, pencils, bathrobe ties or soft rope, items to weigh someone down, and a Bible	HFN, p. 28
Investing and Saving Adds Value to Money	Prov. 21:20	Two and a half dollars for each family member	MMK, p. 87
It Is Important to Spend Time Praising God	Pss. 66:1; 81:1; 95:1; 98:4; 100:1	Plastic straws, scissors, a Bible	SS, p. 52

An Introduction to Family Nights = IFN

Basic Christian Beliefs = BCB

Christian Character Qualities = CCQ

Wisdom Life Skills = WLS

Money Matters for Kids = MMK

Holidays Family Night = HFN

Bible Stories for Preschoolers (Old Testament) = OTS

Simple Science = SS

TOPIC	SCRIPTURE	WHAT YOU'LL NEED	WHERE TO FIND IT
It's Better to Follow the Truth	Rom. 1:25; Prov. 2:1-5	Second set of clues, box of candy or treats, Bible	WLS, p. 86
It's Better to Wait for Something Than to Borrow Money to Buy It	2 Kings 4:1-7; Prov. 22:7	Magazines, advertisements, paper, a pencil, Bible	MMK, p. 103
It's Difficult to Be a Giver When You're a Debtor		Pennies or other coins	MMK, p. 105
It's Easy to Follow a Lie, but It Leads to Disappointment		Clues as described in lesson, empty box	WLS, p. 85
The Importance of Your Name Being Written in the Book of Life	Rev. 20:11-15; 21:27	Bible, phone book, access to other books with family name	BCB, p. 74
It's Important to Listen to Jesus' Message		Bible	BCB, p. 68
It's Not Easy to Break a Pattern of Sin	James 1:12-15	Paper, pan, water, a Bible	SS, p. 63
Jesus Came to Die for Our Sins	Rom. 5:8	A large piece of cardboard, markers, scissors, tape, and a Bible	HFN, p. 91
Jesus Came to Give Us Eternal Life	Mark 16:12-14	A calculator, a calendar, a sheet of paper, and a pencil	HFN, p. 91
Jesus Came to Teach Us about God	John 1:14, 18	Winter clothing, bread crumbs, a Bible	HFN, p. 92
Jesus Came to Show Us How Much God Loves Us	John 3:16	Supplies to make an Advent wreath, and a Bible	HFN, p. 89
Jesus Died for Our Sins	Luke 22:1-6; Mark 14:12-26; Luke 22:47-54; Luke 22:55-62; Matt. 27:1-10; Matt. 27:11-31; Luke 23:26-34	Seven plastic eggs, slips of paper with Scripture verses, and a Bible	HFN, p. 33
Jesus Dies on the Cross	John 14:6	6-foot 2x4, 3-foot 2x4, hammers, nails, Bible	IFN, p. 33
Jesus Promises Us New Bodies and a New Home in Heaven	Phil. 3:20-21; Luke 24:36-43; Rev. 21:1-4	Ingredients for making pumpkin pie, and a Bible	HFN, p. 61

AN INTRODUCTION TO FAMILY NIGHTS = IFN

BASIC CHRISTIAN BELIEFS = BCB

CHRISTIAN CHARACTER QUALITIES = CCQ

WISDOM LIFE SKILLS = WLS

MONEY MATTERS FOR KIDS = MMK

HOLIDAYS FAMILY NIGHT = HFN

BIBLE STORIES FOR PRESCHOOLERS (OLD TESTAMENT) = OTS

SIMPLE SCIENCE = SS

Heritage BUILDERS

AN INTRODUCTION TO FAMILY NIGHTS = IFN

BASIC CHRISTIAN BELIEFS = BCB

CHRISTIAN CHARACTER QUALITIES = CCQ

WISDOM LIFE SKILLS = WLS

MONEY MATTERS FOR KIDS = MMK

HOLIDAYS FAMILY NIGHT = HFN

BIBLE STORIES FOR PRESCHOOLERS (OLD TESTAMENT) = OTS

SIMPLE SCIENCE = SS

TOPIC	SCRIPTURE	WHAT YOU'LL NEED	WHERE TO FIND IT
Jesus Took Our Sins to the Cross and Freed Us from Being Bound Up in Sin	Rom. 6:23, 5:8; 6:18	Soft rope or heavy yarn, a watch with a second hand, thread, and a Bible	HFN, p. 53
Jesus Took the Punishment We Deserve	Rom. 6:23; John 3:16; Rom. 5:8-9	Bathrobe, list of bad deeds	IFN, p. 26
Jesus Was Victorious Over Death and Sin	Luke 23:35-43; Luke 23:44-53; Matt. 27:59-61; Luke 23:54–24:12	Five plastic eggs—four with Scripture verses, and a Bible	HFN, p. 36
Jesus Washes His Followers' Feet	John 13:1-17	Bucket of warm soapy water, towels, Bible	IFN, p. 63
Joshua and the Battle of Jericho	Josh. 1:16-18; 6:1-21	Paper, pencil, dots on paper that, when connected, form a star	IFN, p. 57
Knowing God's Word Helps Us Know What Stand to Take	2 Tim. 3:1-5	Current newspaper, Bible	CCQ, p. 93
Look to God, Not Others	Phil. 4:11-13	Magazines or newspapers, a chair, several pads of small yellow stickies, Bible	WLS, p. 24
Love Is Unselfish	1 Cor. 13	A snack and a Bible	HFN, p. 21
Loving Money Is Wrong	1 Tim. 6:6-10	Several rolls of coins, masking tape, Bible	WLS, p. 45
Lying Can Hurt People	Acts 5:1-11	Two pizza boxes—one empty and one with a fresh pizza—and a Bible	MMK, p. 57
Meeting Goals Requires Planning	Prov. 3:5-6	Paper, scissors, pencils, a treat, a Bible	MMK, p. 71
Moms Are Special and Important to Us and to God	Prov. 24:3-4	Confetti, streamers, a comfortable chair, a wash basin with warm water, two cloths, and a Bible	HFN, p. 41
Moms Model Jesus' Love When They Serve Gladly	2 Tim. 1:4-7	Various objects depending on chosen activity and a Bible	HFN, p. 42
The More We Know God, the More We Know His Voice	John 10:1-6	Bible	BCB, p. 35
Nicodemus Asks Jesus about Being Born Again	John 3:7, 50-51; 19:39-40	Bible, paper, pencil, costume	BCB, p. 81

TOPIC	SCRIPTURE	WHAT YOU'LL NEED	WHERE TO FIND IT
Noah Obeyed God When He Built the Ark	Gen. 6:14-16	A large refrigerator box, markers or paints, self-adhesive paper, stuffed animals, a Bible, utility knife	OTS, p. 23
Nothing Is Impossible When It Is in God's Will	Matt. 21:28	Hard-boiled egg, butter, glass bottle, paper, matches, a Bible	SS, p. 34
Obedience Has Good Rewards		Planned outing everyone will enjoy, directions on 3x5 cards, number cards	IFN, p. 59
Obey God First		Paper, markers, scissors, and blindfolds	OTS, p. 80
Only a Relationship with God Can Fill Our Need	Isa. 55:1-2	Doll that requires batteries, batteries for the doll, dollar bill, pictures of a house, an expensive car, and a pretty woman or handsome man, Bible	WLS, p. 62
Our Actions Should Mirror God, Not the World	Rom. 12:2	Regular glass, dried peas, a wine glass, a pie tin, water, a Bible	SS, p. 57
Our Conscience Helps Us Know Right from Wrong	Rom. 2:14-15	Foods with a strong smell, blindfold, Bible	WLS, p. 93
Our Minds Should Be Filled with Good, Not Evil	Phil 4:8; Ps. 119:9, 11	Bible, bucket of water, several large rocks	CCQ, p. 26
Our Tongue Is Powerful and Should Be Used to Glorify God	James 3:5-8	Squirt gun, pie pan, Pop Rocks candy, a Bible	SS, p. 51
Parable of the Talents	Matt. 25:14-30	Bible	IFN, p. 73
Parable of the Vine and Branches	John 15:1-8	Tree branch, paper, pencils, Bible	IFN, p. 95
People Look at Outside Appearance, but God Looks at the Heart	1 Sam. 17	Slings from activity on p. 73, plastic golf balls or marshmallows, a tape measure, cardboard, markers, and a Bible	OTS, p. 75
Persecution Brings a Reward		Bucket, bag of ice, marker, one-dollar bill	WLS, p. 32
Planning Helps Us Finish Strong	Phil. 3:10-14	Flight map on p. 86, paper, pencils, Bible	CCQ, p. 85

An Introduction to Family Nights
= IFN

Basic Christian Beliefs
= BCB

Christian Character Qualities
= CCQ

Wisdom Life Skills
= WLS

Money Matters for Kids
= MMK

Holidays Family Night
= HFN

Bible Stories for Preschoolers (Old Testament)
= OTS

Simple Science
= SS

Heritage Builders

An Introduction to Family Nights = IFN

Basic Christian Beliefs = BCB

Christian Character Qualities = CCQ

Wisdom Life Skills = WLS

Money Matters for Kids = MMK

Holidays Family Night = HFN

Bible Stories for Preschoolers (Old Testament) = OTS

Simple Science = SS

TOPIC	SCRIPTURE	WHAT YOU'LL NEED	WHERE TO FIND IT
Pray, Endure, and Be Glad When We're Persecuted	Matt. 5:11-12, 44; Rom. 12:14; 1 Cor. 4:12	Notes, Bible, candle or flashlight, dark small space	WLS, p. 29
Putting God First Builds a Solid Relationship	Mark 6:35; Luke 4:16; Mark 13:31; Luke 12:31	Wide-mouth glass jar, large rocks, sand, water, permanent marker, a Bible	SS, p. 70
Remember All God Has Done for You	Ex. 25:1; 16:34; Num. 17:10; Deut. 31:26	Ark of the covenant from p. 49, cardboard or Styrofoam, crackers, a stick, and a Bible	OTS, p. 51
Remember What God Has Done for You	Gen. 12:7-8; 13:18; 22:9	Bricks or large rocks, paint, and a Bible	OTS, p. 31
The Responsibilities of Families	Eph. 5:22-33; 6:1-4	Photo albums, Bible	BCB, p. 101
Satan Looks for Ways to Trap Us	Luke 4:1-13	Cardboard box, string, stick, small ball, Bible	WLS, p. 69
Self-control Helps Us Resist the Enemy	1 Peter 5:8-9; 1 Peter 2:11-12	Blindfold, watch or timer, feather or other "tickly" item, Bible	CCQ, p. 101
Serve One Another in Love	Gal. 5:13	Bag of small candies, at least three per child	IFN, p. 47
Sin and Busyness Interfere with Our Prayers	Luke 10:38-42; Ps. 46:10; Matt. 5:23-24; 1 Peter 3:7	Bible, two paper cups, two paper clips, long length of fishing line	CCQ, p. 61
Sin Separates Humanity	Gen. 3:1-24	Bible, clay creations, piece of hardened clay or play dough	BCB, p. 25
Some Places Aren't Open to Everyone		Book or magazine with "knock-knock" jokes	BCB, p. 73
Some Things in Life Are Out of Our Control		Blindfolds	BCB, p. 41
Sometimes God Surprises Us with Great Things	Gen. 15:15	Large sheet of poster board, straight pins or straightened paper clips, a flashlight, and a Bible	OTS, p. 32
Sometimes We Face Things That Seem Impossible		Bunch of cardboard boxes or blocks	OTS, p. 55
Stand Strong in the Lord	Prov. 1:8-10; 12:3	A jar, string, chair, fan, small weight, a Bible	SS, p. 22

TOPIC	SCRIPTURE	WHAT YOU'LL NEED	WHERE TO FIND IT
Temptation Takes Our Eyes Off God		Fishing pole, items to catch, timer, Bible	IFN, p. 85
Test What the World Offers for Consistency with Jesus' Teachings	1 John 4:1	Candle, apple, almond, a Bible	SS, p. 58
There Is a Difference between Needs and Wants	Prov. 31:16; Matt. 6:21	Paper, pencils, glasses of drinking water, a soft drink	MMK, p. 95
Those Who Don't Believe Are Foolish	Ps. 44:1	Ten small pieces of paper, pencil, Bible	IFN, p. 19
Tithing Means Giving One-Tenth Back to God	Gen. 28:10-22; Ps. 3:9-10	All family members need ten similar items each, a Bible	MMK, p. 33
The Tongue Is Small but Powerful	James 3:3-12	Video, news magazine or picture book showing devastation of fire, match, candle, Bible	IFN, p. 77
The Treasure of a Thankful Heart Is Contentment	Eph. 5:20	3x5 cards, pencils, fun prizes, and a Bible	HFN, p. 72
Trials Help Us Grow	James 1:2-4	Sugar cookie dough, cookie cutters, baking sheets, miscellaneous baking supplies, Bible	WLS, p. 15
Trials Test How We've Grown	James 1:12	Bible	WLS, p. 17
Trust Is Important	Matt. 6:25-34	Each person needs an item he or she greatly values	MMK, p. 25
We All Have Weaknesses and Will Be Attacked by Satan	1 Kings 11:3-4; 2 Cor. 12:9-10	Two pieces of plain white paper, a pencil, a Bible	SS, p. 28
We All Sin	Rom. 3:23	Target and items to throw	IFN, p. 23
We Are a Family for Life, Forever	Ruth 1:4	Shoebox; scissors; paper or cloth; magnets; photos of family members, friends, others; and a Bible	OTS, p. 68
We Are Made in God's Image	Gen. 2:7; Ps. 139:13-16	Paper bags, candies, a Bible, supplies for making gingerbread cookies	HFN, p. 17
We Become a New Creation When Jesus Comes into Our Hearts	Matt. 23:25-28; Rev. 3:20; 2 Cor. 5:17; Eph. 2:10; 2 Cor. 4:7-10; Matt. 5:14-16; 2 Cor. 4:6	Pumpkin, newspaper, sharp knife, a spoon, a candle, matches, and a Bible	HFN, p. 59

An Introduction to Family Nights = IFN

Basic Christian Beliefs = BCB

Christian Character Qualities = CCQ

Wisdom Life Skills = WLS

Money Matters for Kids = MMK

Holidays Family Night = HFN

Bible Stories for Preschoolers (Old Testament) = OTS

Simple Science = SS

TOPIC	SCRIPTURE	WHAT YOU'LL NEED	WHERE TO FIND IT
We Can Communicate with Each Other			BCB, p. 65
We Can Fight the Temptation to Want More Stuff	Matt. 4:1-11; Heb. 13:5	Television, paper, a pencil, Bible	MMK, p. 49
We Can Give Joyfully to Others	Luke 10:25-37	Bible, soft yarn	MMK, p. 41
We Can Help Each Other	Prov. 27:17	Masking tape, bowl of unwrapped candies, rulers, yardsticks, or dowel rods	BCB, p. 110
We Can Help People When We Give Generously	2 Cor. 6–7	Variety of supplies, depending on chosen activity	MMK, p. 43
We Can Learn about God from Mom (or Dad)		Supplies to make a collage (magazines, paper, tape or glue, scissors)	HFN, p. 49
We Can Learn and Grow from Good and Bad Situations	Gen. 37–48; Rom. 8:29	A Bible and a camera (optional)	OTS, p. 37
We Can Love by Helping Those in Need	Heb. 13:1-3		IFN, p. 48
We Can Show Love through Respecting Family Members		Paper and pen	IFN, p. 66
We Can't Hide from God		Supplies will vary	OTS, p. 85
We Can't Take Back the Damage of Our Words		Tube of toothpaste for each child, $10 bill	IFN, p. 78
We Deserve Punishment for Our Sins	Rom. 6:23	Dessert, other materials as decided	IFN, p. 24
We Give to God because We're Thankful		Supplies for a celebration dinner, also money for each family member	MMK, p. 36
We Have All We Need in Our Lives	Ecc. 3:11	Paper, pencils, Bible	WLS, p. 61
We Have a New Life in Christ	John 3:3; 2 Cor. 5:17	Video or picture book of caterpillar forming a cocoon then a butterfly, or a tadpole becoming a frog, or a seed becoming a plant	BCB, p. 93
We Have Much to Be Thankful For	1 Chron. 16:4-36	Unpopped popcorn, a bowl, supplies for popping popcorn, and a Bible	HFN, p. 79

An Introduction to Family Nights = IFN

Basic Christian Beliefs = BCB

Christian Character Qualities = CCQ

Wisdom Life Skills = WLS

Money Matters for Kids = MMK

Holidays Family Night = HFN

Bible Stories for Preschoolers (Old Testament) = OTS

Simple Science = SS

TOPIC	SCRIPTURE	WHAT YOU'LL NEED	WHERE TO FIND IT
We Know Others by Our Relationships with Them		Copies of questionnaire, pencils, Bible	BCB, p. 31
We Must Be in Constant Contact with God		Blindfold	CCQ, p. 63
We Must Choose to Obey		3x5 cards or slips of paper, markers, and tape	IFN, p. 43
We Must Either Choose Christ or Reject Christ	Matt. 12:30	Clear glass jar, cooking oil, water, spoon, Bible	CCQ, p. 96
We Must Give Thanks in All Circumstances	1 Thes. 5:18	A typical family meal, cloth strips, and a Bible	HFN, p. 77
We Must Hold Firm to Our Faith and Depend on God for Strength	Eph. 6:16	Balloons, long darts or shish kebab skewers, cooking oil, a Bible	SS, p. 27
We Must Learn How Much Responsibility We Can Handle		Building blocks, watch with second hand, paper, pencil	IFN, p. 71
We Must Listen	Prov. 1:5, 8-9; 4:1	Bible, other supplies for the task you choose	WLS, p. 77
We Must Think Before We Speak	James 1:19	Bible	WLS, p. 79
We Need to Feed on God's Word to Grow in Christ	Ps. 119:105; 2 Chron. 34:31; Acts 17:11; James 1:22-25	Raisins, clear drinking glass, a two-liter bottle of clear soft drink, a Bible	SS, p. 46
We Need to Grow Closer to Jesus Each Day	Acts 9:1-18	Pitcher, lemonade mix (sugarless), sugar, dry ice, a Bible	SS, p. 45
We Need to Grow Physically, Emotionally, and Spiritually	1 Peter 2:2	Photograph albums or videos of your children at different ages, tape measure, bathroom scale, Bible	CCQ, p. 75
We Prove Who We Are When What We Do Reflects What We Say	James 1:22; 2:14-27	A bag of candy, a rope, and a Bible	HFN, p. 67
We Reap What We Sow	Gal. 6:7	Candy bar, Bible	IFN, p. 55
We Should Do What God Wants Even If We Don't Think We Can		A powerful fan, large sheet of lightweight black plastic, duct tape, and a flashlight	OTS, p. 86

An Introduction to Family Nights = IFN

Basic Christian Beliefs = BCB

Christian Character Qualities = CCQ

Wisdom Life Skills = WLS

Money Matters for Kids = MMK

Holidays Family Night = HFN

Bible Stories for Preschoolers (Old Testament) = OTS

Simple Science = SS

Heritage Builders

An Introduction to Family Nights = IFN
Basic Christian Beliefs = BCB
Christian Character Qualities = CCQ
Wisdom Life Skills = WLS
Money Matters for Kids = MMK
Holidays Family Night = HFN
Bible Stories for Preschoolers (Old Testament) = OTS
Simple Science = SS

TOPIC	SCRIPTURE	WHAT YOU'LL NEED	WHERE TO FIND IT
We Shouldn't Value Possessions Over Everything Else	1 Tim. 6:7-8	Box is optional	CCQ, p. 18
When God Sent Jesus to Earth, God Chose Me	Luke 1:26-38; John 3:16; Matt. 14:23	Going to choose a Christmas tree or other special decoration, a Bible, and hot chocolate	HFN, p. 83
When We Accept Jesus' Gift of Salvation, We Receive the Holy Spirit	John 3:5-8	1/4-full roll of toilet paper, a blow dryer, a dowel rod, a Bible	SS, p. 39
When We Focus on What We Don't Have, We Get Unhappy	1 Tim. 6:9-10; 1 Thes. 5:18; Phil. 4:11-13	A glass, water, paper, crayons, and a Bible	HFN, p. 71
When We're Set Free from Sin, We Have the Freedom to Choose, and the Responsibility to Serve	Gal. 5:13-15	Candies, soft rope, and a Bible	HFN, p. 55
Wise Spending Means Getting Good Value for What We Buy	Luke 15:11-32	Money and a Bible	MMK, p. 97
With Help, Life Is a Lot Easier		Supplies to do the chore you choose	BCB, p. 101
Wolves in Sheeps' Clothing	Matt. 7:15-20	Ten paper sacks, a marker, ten small items, Bible	IFN, p. 97
Worrying Doesn't Change Anything		Board, inexpensive doorbell buzzer, a 9-volt battery, extra length of electrical wire, a large belt, assorted tools	CCQ, p. 37
You Look Like the Person in Whose Image You Are Created		Paper roll, crayons, markers, pictures of your kids and of yourself as a child	BCB, p. 23

Welcome to the Family!

We hope you've enjoyed this book. Heritage Builders was founded in 1995 by three fathers with a passion for the next generation. As a new ministry of Focus on the Family, Heritage Builders strives to equip, train and motivate parents to become intentional about building a strong spiritual heritage.

It's quite a challenge for busy parents to find ways to build a spiritual foundation for their families—especially in a way they enjoy and understand. Through activities and participation, children can learn biblical truth in a way they can understand, enjoy—and *remember*.

Passing along a heritage of Christian faith to your family is a parent's highest calling. Heritage Builders' goal is to encourage and empower you in this great mission with practical resources and inspiring ideas that really work—and help your children develop a lasting love for God.

***

How To Reach Us

For more information, visit our Heritage Builders Web site! Log on to **www.heritagebuilders.com** to discover new resources, sample activities, and ideas to help you pass on a spiritual heritage. To request any of these resources, simply call Focus on the Family at 1-800-A-FAMILY (1-800-232-6459) or in Canada, call 1-800-661-9800. Or send your request to Focus on the Family, Colorado Springs, CO 80995.
In Canada, write Focus on the Family, P.O. Box 9800, Stn. Terminal, Vancouver, B.C. V6B 4G3

To learn more about Focus on the Family or to find out if there is an associate office in your country, please visit www. family.org

We'd love to hear from you!

Try These Heritage Builders Resources!

Parents' Guide to the Spiritual Growth of Children

Building a foundation of faith in your children can be easy–and fun!–with help from the ***Parents' Guide to the Spiritual Growth of Children***. Through simple and practical advice, this comprehensive guide shows you how to build a spiritual training plan for your family and it explains what to teach your children at different ages.

Bedtime Blessings

Strengthen the precious bond between you, your child and God by making ***Bedtime Blessings*** a special part of your evenings together. From best-selling author John Trent, Ph.D., and Heritage Builders, this book is filled with stories, activities and blessing prayers to help you practice the biblical model of "blessing."

My Time With God

Send your child on an amazing adventure—a self-guided tour through God's Word! ***My Time With God*** shows your 8- to 12-year-old how to get to know God regularly in exciting ways. Through 150 days' worth of fun facts and mind-boggling trivia, prayer starters, and interesting questions, your child will discover how awesome God really is!

The Singing Bible

Children ages 2 to 7 will love ***The Singing Bible***, which sets the Bible to music with over 50 fun, sing-along songs! Lead your child through Scripture by using ***The Singing Bible*** to introduce the story of Jonah, the Ten Commandments and more. This is a fun, fast-paced journey kids will remember.

. . .

Visit our Heritage Builders Web Site! Log on to **www.heritagebuilders. com** to discover new resources, sample activities, and ideas to help you pass on a spiritual heritage. To request any of these resources, simply call Focus on the Family at 1-800-A-FAMILY (1-800-232-6459) or in Canada, call 1-800-661-9800. Or send your request to Focus on the Family, Colorado Springs, CO 80995. In Canada, Write Focus on the Family, P.O. Box 9800, Stn. Terminal, Vancouver, B.C. V6B 4G3.

Every family has a heritage—a spiritual, emotional, and social legacy passed from one generation to the next. There are four main areas we at Heritage Builders recommend parents consider as they plan to pass their faith to their children:

Family Fragrance

Every family's home has a fragrance. Heritage Builders encourages parents to create a home environment that fosters a sweet, Christ-centered AROMA of love through Affection, Respect, Order, Merriment, and Affirmation.

Family Traditions

Whether you pass down stories, beliefs and/or customs, traditions can help you establish a special identity for your family. Heritage Builders encourages parents to set special "milestones" for their children to help guide them and move them through their spiritual development.

Family Compass

Parents have the unique task of setting standards for normal, healthy living through their attitudes, actions and beliefs. Heritage Builders encourages parents to give their children the moral navigation tools they need to succeed on the roads of life.

Family Moments

Creating special, teachable moments with their children is one of a parent's most precious and sometimes, most difficult responsibilities. Heritage Builders encourages parents to capture little moments throughout the day to teach and impress values, beliefs, and biblical principles onto their children.

We look forward to standing alongside you as you seek to impart the Lord's care and wisdom onto the next generation—onto your children.